Undergraduate Texts in Mathematics

Editors

S. Axler
F.W. Gehring
K.A. Ribet

Springer
New York
Berlin
Heidelberg
Hong Kong
London
Milan
Paris
Tokyo

Undergraduate Texts in Mathematics

(continued after index)

John Stillwell

Elements of Number Theory

With 35 figures

 Springer

John Stillwell
Mathematics Department
University of San Francisco
2130 Fulton Street
San Francisco, CA 94117-1080
USA
stillwell@usfca.edu

Mathematics Subject Classification (2000): 11-01

Library of Congress Cataloging-in-Publication Data
Stillwell, John.
    Elements of number theory / John Stillwell.
       p. cm. — (Undergraduate texts in mathematics)
    Includes bibliographical references and index.
       ISBN 0-387-95587-9. I. Title. II. Series.
QA241 .S813 2002
512'.7—dc21   2002030569

ISBN 0-387-95587-9   Printed on acid-free paper.

Printed in the United States of America.

9 8 7 6 5 4 3 2 1   SPIN 10892865

www.springer-ny.com

Springer-Verlag   New York Berlin Heidelberg
A member of BertelsmannSpringer Science+Business Media GmbH

To Elaine

Preface

This book is intended to complement my *Elements of Algebra*, and it is similarly motivated by the problem of solving polynomial equations. However, it is independent of the algebra book, and probably easier. In *Elements of Algebra* we sought *solution by radicals*, and this led to the concepts of *fields* and *groups* and their fusion in the celebrated theory of Galois. In the present book we seek *integer solutions*, and this leads to the concepts of *rings* and *ideals* which merge in the equally celebrated *theory of ideals* due to Kummer and Dedekind.

Solving equations in integers is the central problem of number theory, so this book is truly a number theory book, with most of the results found in standard number theory courses. However, numbers are best understood through their algebraic structure, and the necessary algebraic concepts— rings and ideals—have no better motivation than number theory.

The first nontrivial examples of rings appear in the number theory of Euler and Gauss. The concept of ideal—today as routine in ring theory as the concept of normal subgroup is in group theory—also emerged from number theory, and in quite heroic fashion. Faced with failure of unique prime factorization in the arithmetic of certain generalized "integers", Kummer created in the 1840s a new kind of number to overcome the difficulty. He called them "ideal numbers" because he did not know exactly what they were, though he knew how they behaved. Dedekind in 1871 found that these "ideal numbers" could be realized as *sets* of actual numbers, and he called these sets *ideals*.

Dedekind found that ideals could be defined quite simply; so much so that a student meeting the concept today might wonder what all the fuss is about. It is only in their role as "ideal numbers", where they realize Kummer's impossible dream, that ideals can be appreciated as a genuinely brilliant idea.

Thus solution in integers—like solution by radicals—is a superb setting in which to show algebra at its best. It is the right place to introduce rings and ideals and the right place first to apply them. It even gives an opportunity to introduce some exotic rings, such as the quaternions, which we use to prove Lagrange's theorem that every natural number is the sum of four squares.

The book is based on two short courses (about 20 lectures each) given at Monash University in recent years; one on elementary number theory and one on ring theory with applications to algebraic number theory. Thus the amount of material is suitable for a one-semester course, with some variation possible through omission of the optional starred sections. A slower-paced course could stop at the end of Chapter 9, at which point most of the standard results have been covered, from Euclid's theorem that there are infinitely many primes to quadratic reciprocity.

It should be stressed, however, that this is not meant to be a standard number theory course. I have tried to avoid the ad hoc proofs that once gave number theory a bad name, in favor of unifying ideas that work in many situations. These include algebraic structures but also ideas from elementary number theory, such as the Euclidean algorithm and unique prime factorization. In particular, I use the Euclidean algorithm as a bridge to Conway's visual theory of quadratic forms, which offers a new approach to the Pell equation.

There are exercises at the end of almost every section, so that each new idea or proof receives immediate reinforcement. Some of them focus on specific ideas, while others recapitulate the general line of argument (in easy steps) to prove a similar result. The purpose of each exercise should be clear from the accompanying commentary, so instructors and independent readers alike will be able to find an enjoyable path through the book.

My thanks go to the Monash students who took the courses on which the book is based. Their reactions have helped improve the presentation in many ways. I am particularly grateful to Ley Wilson, who showed that it is possible to master the book by independent study.

Special thanks go to my wife Elaine, who proofread the first version of the book, and to John Miller and Abe Shenitzer, who carefully read the revised version and saved me from many mathematical and stylistic errors.

JOHN STILLWELL
South Melbourne, July 2002

Contents

1

Natural numbers and integers

PREVIEW

Counting is presumably the origin of mathematical thought, and it is certainly the origin of difficult mathematical problems. As the great Hungarian problem-solver Paul Erdős liked to point out, if you can think of an open problem that is more than 200 years old, then it is probably a problem in number theory.

In recent decades, difficulties in number theory have actually become a virtue. *Public key encryption*, whose security depends on the difficulty of factoring large numbers, has become one of the commonest applications of mathematics in daily life.

At any rate, problems are the life blood of number theory, and the subject advances by building theories to make them understandable. In the present chapter we introduce some (not so difficult) problems that have played an important role in the development of number theory because they lead to basic methods and concepts.

- Counting leads to *induction*, the key to all facts about numbers, from banalities such as $a + b = b + a$ to the astonishing result of Euclid that there are infinitely many primes.

- Division (with remainder) is the key computational tool in Euclid's proof and elsewhere in the study of primes.

- Binary notation, which also results from division with remainder, leads in turn to a method of "fast exponentiation" used in public key encryption.

- The Pythagorean equation $x^2 + y^2 = z^2$ from geometry is equally important in number theory because it has integer solutions.

1

In this chapter we are content to show these ideas at work in few
interesting but seemingly random situations. Later chapters will develop the ideas in more depth, showing how they unify and explain
a great many astonishing properties of numbers.

1.1 Natural numbers

Number theory starts with the *natural numbers*

$$1,2,3,4,5,6,7,8,9,\ldots,$$

generated from 1 by successively adding 1. We denote the set of natural
numbers by $\mathbb{N}$. On $\mathbb{N}$ we have the operations $+$ and $\times$, which are simple in
themselves but lead to more sophisticated concepts.

For example, we say that *a divides n* if $n = ab$ for some natural numbers
a and b. A natural number p is called *prime* if the only natural numbers
dividing p are 1 and p itself.

Divisibility and primes are behind many of the interesting questions in
mathematics, and also behind the recent applications of number theory (in
cryptography, internet security, electronic money transfers etc.).

The sequence of prime numbers begins with

$$2,3,5,7,11,13,17,19,23,29,31,37,\ldots$$

and continues in a seemingly random manner. There is so little pattern in
the sequence that one cannot even see clearly whether it continues forever.
However, Euclid (around 300 BCE) proved that *there are infinitely many
primes*, essentially as follows.

Infinitude of primes. *Given any primes $p_1, p_2, p_3, \ldots, p_k$, we can always
find another prime p.*

Proof. Form the number

$$N = p_1 p_2 p_3 \cdots p_k + 1.$$

Then none of the given primes $p_1, p_2, p_3, \ldots, p_k$ divides N because they all
leave remainder 1. On the other hand, *some* prime p divides N. If N itself
is prime we can take $p = N$, otherwise $N = ab$ for some smaller numbers a
and b. Likewise, if either a or b is prime we take it to be p, otherwise split
a and b into smaller factors, and so on. Eventually we must reach a prime
p dividing N because natural numbers cannot decrease forever. $\square$

Exercises

Not only is the sequence of primes without apparent pattern, there is not even a known simple formula that produces only primes. There are, however, some interesting "near misses".

1.1.1 Check that the quadratic function $n^2 + n + 41$ is prime for all small values of n (say, for n up to 30).

1.1.2 Show nevertheless that $n^2 + n + 41$ is not prime for certain values of n.

1.1.3 Which is the smallest such value?

1.2   Induction

The method just used to find the prime divisors of N is sometimes called *descent*, and it is an instance of a general method called *induction*.

The "descent" style of induction argument relies on the fact that any process producing smaller and smaller natural numbers must eventually halt. The process of repeatedly adding 1 reaches any natural number n in a finite number of steps, hence there are only finitely many steps *downward* from n. There is also an "ascent" style of induction that imitates the construction of the natural numbers themselves—starting at some number and repeatedly adding 1.

An "ascent" induction proof is carried out in two steps: the *base step* (getting started) and the *induction step* (going from n to $n + 1$). Here is an example: proving that *any number of the form $k^3 + 2k$ is divisible by* 3.

Base step. The claim is true for $k = 1$ because $1^3 + 2 \times 1 = 3$, which is certainly divisible by 3.

Induction step. Suppose that the claim is true for $k = n$, that is, 3 divides $n^3 + 2n$. We want to deduce that it is true for $k = n + 1$, that is, that 3 divides $(n + 1)^3 + 2(n + 1)$. Well,

$$
\begin{aligned}
&(n + 1)^3 + 2(n + 1) \\
&= n^3 + 3n^2 + 3n + 1 + 2n + 2 \\
&= n^3 + 2n + 3n^2 + 3n + 3 \\
&= n^3 + 2n + 3(n^2 + n + 1)
\end{aligned}
$$

And the right-hand side is the sum of $n^3 + 2n$, which we are supposing to be divisible by 3, and $3(n^2 + n + 1)$, which is obviously divisible by 3. Therefore $(n + 1)^3 + 2(n + 1)$ is divisible by 3, as required.   □

Induction is fundamental not only for proofs of theorems about $\mathbb{N}$ but also for defining the basic functions on $\mathbb{N}$. Only one function needs to be assumed, namely the *successor function* $s(n) = n + 1$; then $+$ and $\times$ can be defined by induction. In this book we are not trying to build everything up from bedrock, so we shall assume $+$ and $\times$ and their basic properties, but it is worth mentioning their inductive definitions, since they are so simple.

For any natural number m we define $m + 1$ by

$$m + 1 = s(m).$$

Then, given the definition of $m + n$ for all m, we define $m + s(n)$ by

$$m + s(n) = s(m + n).$$

It then follows, by induction on n, that $m + n$ is defined for all natural numbers m and n. The definition of $m \times n$ is similarly based on the successor function and the $+$ function just defined:

$$m \times 1 = m$$
$$m \times s(n) = m \times n + m.$$

From these inductive definitions one can give inductive *proofs* of the basic properties of $+$ and $\times$, for example $m + n = n + m$ and $l(m + n) = lm + ln$. Such proofs were first given by Grassmann (1861) (in a book intended for high school students!) but they went unnoticed. They were rediscovered, together with an analysis of the successor function itself, by Dedekind (1888). For more on this see Stillwell (1998), Chapter 1.

Exercises

An interesting process of descent may be seen in the algorithm for the so-called *Egyptian fractions* introduced by Fibonacci (1202). The goal of the algorithm is to represent any fraction $\frac{b}{a}$ with $0 < b < a$ as sum of distinct terms $\frac{1}{n}$, called *unit fractions*. (The ancient Egyptians represented fractions in this way.)

Fibonacci's algorithm, in a nutshell, is to *repeatedly subtract the largest possible unit fraction*. Applied to the fraction $\frac{11}{12}$, for example, it yields

$$\frac{11}{12} - \frac{1}{2} = \frac{5}{12}, \quad \text{subtracting the largest unit fraction, } \tfrac{1}{2}, \text{ less than } \tfrac{11}{12},$$

$$\frac{5}{12} - \frac{1}{3} = \frac{1}{12}, \quad \text{subtracting the largest unit fraction, } \tfrac{1}{3}, \text{ less than } \tfrac{5}{12},$$

$$\text{hence } \frac{11}{12} = \frac{1}{2} + \frac{1}{3} + \frac{1}{12}.$$

It turns out that the fractions produced by the successive subtractions always have a descending sequence of numerators (11, 5, 1 in the example), hence they necessarily terminate with 1.

1.2.1 Use Fibonacci's algorithm to find an Egyptian representation of $\frac{9}{11}$.

1.2.2 If a, b, q are natural numbers with $\frac{1}{q+1} < \frac{b}{a} < \frac{1}{q}$, show that

$$\frac{b}{a} - \frac{1}{q+1} = \frac{b'}{a(q+1)} \quad \text{where} \quad 0 < b' < b.$$

Hence explain why Fibonacci's algorithm always works.

1.3   Integers

For several reasons, it is convenient to extend the set $\mathbb{N}$ of natural numbers to the *group* $\mathbb{Z}$ of *integers* by throwing in the *identity* element 0 and an *inverse* $-n$ for each natural number n. One reason for doing this is to ensure that the difference $m - n$ of any two integers is meaningful. Thus $\mathbb{Z}$ is a set on which all three operations $+$, $-$, and $\times$ are defined. (The notation $\mathbb{Z}$ comes from the German "Zahlen", meaning "numbers".)

$\mathbb{Z}$ is an *abelian group* under the operation $+$, because it has the three group properties:

Associativity: $a + (b + c) = (a + b) + c$
Identity: $a + 0 = a$
Inverse: $a + (-a) = 0$

and also the abelian property: $a + b = b + a$.

$\mathbb{Z}$ is much older than the concept of abelian group. The latter concept could only be conceived after other examples came to light, particularly *finite* abelian groups. We shall meet some of them in Chapter 3.

$\mathbb{Z}$ is a *ring* under the operations $+$ and $\times$: it is an abelian group under $+$ and the $\times$ is linked with $+$ by

Distributivity: $a(b + c) = ab + ac$.

The ring concept also emerged much later than $\mathbb{Z}$. It grew out of 18th and 19th century attempts to generalize the concept of integer. We see one of these in Section 1.8, and take up the general ring concept in Chapter 10.

The ring properties show that $\mathbb{Z}$ has more structure than $\mathbb{N}$, though it must be admitted that this does not make everything simpler. The presence of the negative integers $-1, -2, -3, \ldots$ in $\mathbb{Z}$ slightly complicates the concept of prime number. Since any integer n is divisible by $1, -1, n$ and $-n$, we have to define a *prime* in $\mathbb{Z}$ to be an integer p divisible only by ± 1 (the so-called *units* of $\mathbb{Z}$) and $\pm p$.

In general, however, it is simpler to work with integers than natural numbers. Here is a problem that illustrates the difference.

Problem. Describe the numbers $4m + 7n$

1. where m and n are natural numbers,

2. where m and n are integers.

In the first case the numbers are 11, 15, 18, 19, 22, 23, 25, 26, 27 and all numbers ≥ 29. The numbers < 29 can be verified (laboriously, I admit) by trial. To see why all numbers ≥ 29 are of the form $4m + 7m$, we first verify this for 29, 30, 31, 32; namely

$$29 = 2 \times 4 + 3 \times 7$$
$$30 = 4 \times 4 + 2 \times 7$$
$$31 = 6 \times 4 + 1 \times 7$$
$$32 = 1 \times 4 + 4 \times 7.$$

Then we can get the next four natural numbers by adding one more 4 to each of these, then the next four by adding two more 4s, and so on (this is really an induction proof).

In the second case, all integers are obtainable. This is simply because $1 = 4 \times 2 - 7$, and therefore $n = 4 \times 2n - 7 \times n$, for any integer n.

This type of problem is easier to understand with the help of the gcd—*greatest common divisor*—which we study in the next chapter. But first we need to look more closely at division, particularly division with remainder, which is the subject of the next section.

Exercises

A concrete problem similar to describing $4m + 7n$ is the *McN*ggets problem*: given that McN*ggets can be bought in quantities of 6, 9 or 20, which numbers of McN*ggets can be bought? This is the problem of describing the numbers $6i + 9j + 20k$ for natural numbers or zero i, j and k.

It turns out the possible numbers include all numbers ≥ 44, and an irregular set of numbers < 43.

1.3.1 Explain why the number 43 is not obtainable.

1.3.2 Show how each of the numbers 44, 45, 46, 47, 48, 49 is obtainable.

1.3.3 Deduce from Exercise 1.3.2 that any number > 43 is obtainable.

But if the negative quantities −6, −9 and −20 are allowed (say, by selling McN*ggets back), then any integer number of McN*ggets can be obtained.

1.3.4 Show in fact that $1 = 9m + 20n$ for some integers m and n.

1.3.5 Deduce from Exercise 1.3.4 that every integer is expressible in the form $9m + 20n$, for some integers m and n.

1.3.6 Is every integer expressible in the form $6m + 9n$? What do the results in Exercises 1.3.4 and 1.3.5 have to do with common divisors?

1.4   Division with remainder

As mentioned in Section 1.1, a natural number b is said to *divide n* if $n = bc$ for some natural number c. We also say that b is a *divisor of n*, and that n is a *multiple of b*. The same definitions apply wherever there is a concept of multiplication, such as in $\mathbb{Z}$.

In $\mathbb{N}$ or $\mathbb{Z}$ it may very well happen that b does *not* divide a, for example, 4 does not divide 23. In this case we are interested in the *quotient q* and *remainder r* when we do *division of a by b*. The quotient comes from the greatest multiple qb of b that is $\leq a$, and the remainder is $a - qb$. For example

$$23 = 5 \times 4 + 3,$$

so when we divide 23 by 4 we get quotient 5 and remainder 3.

The remainder $r = a - qb$ may be found by repeatedly subtracting b from a. This gives natural numbers $a, a - b, a - 2b, \ldots$, which decrease and therefore include least member $r = a - qb \geq 0$ by descent. Then $r < b$, otherwise we could subtract b again. The remainder $r < b$ is also evident in Figure 1.1, which shows a lying between successive multiples of b, hence necessarily at distance $< b$ from the nearest such multiple, qb.

Figure 1.1: Division with remainder

Important. The main purpose of division with remainder is to find the remainder, which tells us whether b divides a or not.

It does not help (and it may be confusing) to form the fraction a/b, because this brings us no closer to knowing whether b divides a. For example, the fraction

$$\frac{43560029}{77777}$$

does not tell us whether 77777 divides 43560039 or not. To find out, we need to know whether the remainder is 0 or not. We could do the full division with remainder:

$$43560029 = 560 \times 77777 + 4909$$

which tells us the exact remainder, 4909, or else evaluate the fraction numerically

$$\frac{43560029}{77777} = 560.0631...,$$

which is enough to tell us that the remainder is $\neq 0$. (And we can read off the quotient $q = 560$ as the part before the decimal point, and hence find the remainder, as $43560029 - 560 \times 77777 = 4909$.)

Exercises

1.4.1 Using a calculator or computer, use the method above to find the remainder when 12345678 is divided by 3333.

1.4.2 Calculate the multiples of 3333 on either side of 12345678.

1.5   Binary notation

Division with remainder is the natural way to find the *binary numeral* of any natural number n. The digits of the numeral are found by dividing n by 2, writing the remainder, and repeating the process with the quotient until the quotient 0 is obtained. Then the sequence of remainders, written in reverse order, is the binary numeral for n.

Example. Binary numeral for 2001.

$$2001 = 1000 \times 2 + 1$$
$$1000 = 500 \times 2 + 0$$
$$500 = 250 \times 2 + 0$$
$$250 = 125 \times 2 + 0$$
$$125 = 62 \times 2 + 1$$
$$62 = 31 \times 2 + 0$$
$$31 = 15 \times 2 + 1$$
$$15 = 7 \times 2 + 1$$
$$7 = 3 \times 2 + 1$$
$$3 = 1 \times 2 + 1$$
$$1 = 0 \times 2 + 1.$$

Hence the binary numeral for 2001 is 11111010001.

A general binary numeral $a_k a_{k-1} \ldots a_1 a_0$, where each a_i is 0 or 1, stands for the number

$$n = a_k 2^k + a_{k-1} 2^{k-1} + \cdots + a_1 2 + a_0,$$

because repeated division of this number by 2 yields the successive remainders $a_0, a_1, \ldots, a_{k-1}, a_k$. Thus one can reconstruct n from its binary digits by multiplying them by the appropriate powers of 2 and adding.

However, it is more efficient to view $a_k a_{k-1} \ldots a_1 a_0$ as a code for constructing n from the number 0 by a sequence of doublings (multiplications by 2) and additions of 1, namely the *reverse of the sequence of operations by which the binary numeral was computed from n*. Moving from left to right, one doubles and adds a_i (if nonzero) for each digit a_i.

Figure 1.2 shows a way to set out the computation, recovering 2001 from its binary numeral 11111010001.

The number of operations

The number of doublings in this process is one less than the number of digits in the binary numeral for n, hence less than $\log_2 n$, since the largest number with k digits is $2^k - 1$ (whose binary numeral consists of k ones), and its log to base 2 is therefore $< k = \log_2(2^k)$.

1	1	1	1	1	0	1	0	0	0	1			
+1												=	1
× 2												=	2
	+1											=	3
	×2											=	6
		+1										=	7
		×2										=	14
			+1									=	15
			×2									=	30
				+1								=	31
				×2								=	62
					+0							=	62
					×2							=	124
						+1						=	125
						×2						=	250
							+0					=	250
							×2					=	500
								+0				=	500
								×2				=	1000
									+0			=	1000
									×2			=	2000
										+1		=	2001

Figure 1.2: Recovering a number from its binary numeral

Likewise, there are $< \log_2 n$ additions. So the total number of operations, either doubling or adding 1, needed to produce n is $< 2\log_2 n$.

This observation gives a highly efficient way to compute powers, based on *repeated squaring*. To form m^n, we begin with $m = m^1$, and repeatedly double the exponent (by squaring) or add 1 to it (by multiplying by m). Since we can reach exponent n by doubling or adding 1 less than $2\log_2 n$ times, we can form m^n by squaring or multiplying by m less than $2\log_2 n$ times. That is, *it takes less than $2\log_2 n$ multiplications to form m^n*.

Thus the number of operations is roughly proportional to the length of n (the number of its binary or decimal digits). Few problems in number theory can be solved in so few steps, and the fast solution of this particular problem is crucial in modern cryptography and electronic security systems (see Chapter 4).

Exercises

Binary notation is more often used by computers than humans, since we have 10 fingers and hence find it convenient to use base 10 rather than base 2. However, some famous numbers are most simply written in binary. Examples are the *Mersenne primes*, which are prime numbers of the form $2^p - 1$ where p is prime.

1.5.1 Show that the binary numeral for $2^p - 1$ is $111 \cdots 1$ (p digits), and that the first four Mersenne primes have binary numerals 11, 111, 11111, and 1111111.

1.5.2 However, not every prime p gives a prime $2^p - 1$: factorize $2^{11} - 1$.

1.5.3 Show also that $2^n - 1$ is *never* a prime when n is not prime. (*Hint*: suppose that $n = pq$, let $x = 2^p$, and show that $x - 1$ divides $x^q - 1$.)

Mersenne primes are named after Marin Mersenne (1588–1648) who first drew attention to the problem of finding them. They occur (though not under that name) in a famous theorem of Euclid on *perfect numbers*. A number is called perfect if it equals the sum of its proper divisors (divisors less than itself). For example, 6 is perfect, because its proper divisors are 1, 2 and 3, and $6 = 1 + 2 + 3$. Euclid's theorem is: *if $2^p - 1$ is prime then $2^{p-1}(2^p - 1)$ is perfect*.

We discuss this theorem further in Chapter 2 when we have developed some theory of divisibility. In the meantime we observe that Euclid's perfect numbers also have binary numerals of a simple form.

1.5.4 Show that the first four perfect numbers arising from Mersenne primes have binary numerals 110, 11100, 111110000, and 1111111000000.

1.5.5 What is the binary numeral for $2^{p-1}(2^p - 1)$?

1.6 Diophantine equations

Solving equations is the traditional goal of algebra, and particular parts of algebra have been developed to analyze particular methods of solution. *Solution by radicals* is one branch of the tradition, typified by the ancient formula

$$x = \frac{-b \pm \sqrt{b^2 - 4ac}}{2a}$$

for the solution of the general quadratic equation $ax^2 + bx + c = 0$, and by more complicated formulas (involving cube roots as well as square roots) for the solution of cubic and quartic equations. This method of solution is analyzed by means of the *field* and *group* concepts, which lead to *Galois theory*. Its main results may be found in the companion book to this one, Stillwell (1994).

The other important branch of the tradition is *finding integer solutions*, the main theme of the present book. It leads to the *ring* concept and *ideal theory*. Equations whose integer solutions are sought are called *Diophantine*, even though it is not really the equations that are "Diophantine", but the solutions. Nevertheless, certain equations stand out as "Diophantine" because their integer solutions are of exceptional interest.

- The Pythagorean equation $x^2 + y^2 = z^2$, whose natural number solutions (x, y, z) are known as *Pythagorean triples*.

- The Pell equation $x^2 - ny^2 = 1$ for any nonsquare natural number n.

- The Bachet equation $y^3 = x^2 + n$ for any natural number n.

- The Fermat equation $x^n + y^n = z^n$ for any integer $n > 2$.

The Pythagorean equation is the oldest known mathematical problem, being the subject of a Babylonian clay tablet from around 1800 BCE known as Plimpton 322 (from its museum catalogue number). The tablet contains the two columns of natural numbers, y and z shown in Figure 1.3.

y	z
119	169
3367	4825
4601	6649
12709	18541
65	97
319	481
2291	3541
799	1249
481	769
4961	8161
45	75
1679	2929
161	289
1771	3229
56	106

Figure 1.3: Plimpton 322

The left part of the table is missing, but it is surely a column of values of x, because each value of $z^2 - y^2$ is an integer square x^2, and so the table is essentially a list of Pythagorean triples.

This means that Pythagorean triples were known long before Pythagoras (who lived around 500 BCE), and the Babylonians apparently had sophisticated means of producing them. Notice that Plimpton 322 does not contain any well known Pythagorean triples, such as $(3, 4, 5)$, $(5, 12, 13)$ or $(8, 15, 17)$. It does, however, contain triples derived from these, mostly in nontrivial ways.

Around 300 BCE, Euclid showed that all natural number solutions of $x^2 + y^2 = z^2$ can be produced by the formulas

$$x = (u^2 - v^2)w, \quad y = 2uvw, \quad z = (u^2 + v^2)w$$

by letting u, v and w run through all the natural numbers. (Also the same formulas with x and y interchanged.)

It is easily checked that these formulas give

$$x^2 + y^2 = z^2,$$

but it is not so easily seen that every solution is of Euclid's form. Another approach, using rational numbers, was found by Diophantus around 200 CE. Diophantus specialized in solving equations in rationals, so his solutions are not properly "Diophantine" in our sense, but in this case rational and integer solutions are essentially equivalent.

Exercises

1.6.1 Check (preferably with the help of computer) that $z^2 - y^2$ is a perfect square for each pair (y, z) in Plimpton 322.

1.6.2 Check also that x is a "round" number in the Babylonian sense, that is generally divisible by 60, or at least by a divisor of 60. (The Babylonian system of numerals had base 60.)

1.6.3 Verify that if

$$x = (u^2 - v^2)w, \quad y = 2uvw, \quad z = (u^2 + v^2)w$$

then $x^2 + y^2 = z^2$.

1.6.4 Find values of u and v (with $w = 1$) that yield the Pythagorean triples $(3,4,5)$, $(5,12,13)$, $(7,24,25)$ and $(8,15,17)$ when substituted in Euclid's formulas.

1.7   The Diophantus chord method

An integer solution $(x,y,z) = (a,b,c)$ of $x^2 + y^2 = z^2$ implies

$$\left(\frac{a}{c}\right)^2 + \left(\frac{b}{c}\right)^2 = 1,$$

so $X = a/c$, $Y = b/c$ is a *rational* solution of the equation

$$X^2 + Y^2 = 1,$$

in other words, a *rational point* on the unit circle. (Admittedly, any multiple of the triple, (ma, mb, mc), corresponds to the same point, but we can easily insert multiples once we have found a, b and c from X and Y.)

Diophantus found rational points on $X^2 + Y^2 = 1$ by an algebraic method, which has the geometric interpretation shown in Figure 1.4.

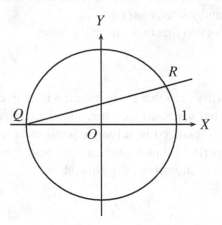

Figure 1.4: The chord method for rational points

If we draw the chord connecting an arbitrary rational point R to the point $Q = (-1, 0)$ we get a line with rational slope, because the coordinates of R and Q are rational. If the slope is t, the equation of this line is

$$Y = t(X + 1).$$

Conversely, any line of this form, with rational slope t, meets the circle at a rational point R. This can be seen by computing the coordinates of R. We do this by substituting $Y = t(X + 1)$ in $X^2 + Y^2 = 1$, obtaining

$$X^2 + t^2(X + 1)^2 = 1,$$

which is the following quadratic equation for X:

$$X^2(1+t^2)+2t^2X+t^2-1=0.$$

The quadratic formula gives the solutions

$$X=-1,\frac{1-t^2}{1+t^2}.$$

The solution $X=-1$ corresponds to the point Q, so the X coordinate at R is $\frac{1-t^2}{1+t^2}$, and hence the Y coordinate is

$$Y=t\left(\frac{1-t^2}{1+t^2}+1\right)=\frac{2t}{1+t^2}.$$

To sum up: an arbitrary rational point on the unit circle $X^2+Y^2=1$ has coordinates

$$\left(\frac{1-t^2}{1+t^2},\frac{2t}{1+t^2}\right),\quad\text{for arbitrary rational }t.$$

Now we can recover Euclid's formulas.

An arbitrary rational t can be written $t=v/u$ where $u,v\in\mathbb{Z}$, and the rational point R then becomes

$$\left(\frac{1-\frac{v^2}{u^2}}{1+\frac{v^2}{u^2}},\frac{2\frac{v}{u}}{1+\frac{v^2}{u^2}}\right)=\left(\frac{u^2-v^2}{u^2+v^2},\frac{2uv}{u^2+v^2}\right).$$

Thus if this is

$$\left(\frac{x}{z},\frac{y}{z}\right)\quad\text{for some }x,y,z\in\mathbb{Z}$$

we must have

$$\frac{x}{z}=\frac{u^2-v^2}{u^2+v^2},\quad\frac{y}{z}=\frac{2uv}{u^2+v^2}$$

for some $u,v\in\mathbb{Z}$.

Euclid's formulas for x, y and z also give these formulas for x/z and y/z, so the results of Euclid and Diophantus are essentially the same.

There is little difference between rational and integer solutions of the equation $x^2+y^2=z^2$ because it is *homogeneous* in x, y and z, hence any rational solution can be multiplied through to give an integer solution. The situation is quite different with inhomogeneous equations, such as $y^2 = x^3 - 2$, where the integer solutions may be much harder to find.

Diophantus' method for rational solutions can be generalized to cubic equations, where it has enjoyed great success. See for example, Silverman and Tate (1992). However, it does not yield integer solutions except in the rare cases where the equation is homogeneous, and hence it diverges from the path we follow in this book. Indeed, it is often the case—for example with Bachet equations—that a cubic equation has infinitely many rational solutions and only finitely many integer solutions. Since we wish to study integer solutions, we now take our leave of the chord construction, and turn in the next section to an algebraic approach to Pythagorean triples: the use of "generalized integers".

Exercises

Diophantus himself extended his method to equations of the form

$$y^2 = \text{cubic in } x,$$

where all coefficients are rational. Here the link between the geometry and the algebra is that *a straight line through two rational points meets the curve in a third rational point*. When there is only one "obvious" rational point on the curve, then one can use the tangent through this point instead of a chord, because the tangent meets the curve twice when viewed algebraically.

The equation $y^2 = x^3 - 2$ is a good one to illustrate the tangent method, as well as the formidable calculations it can lead to. (Note that this is a Bachet equation; here we have interchanged x and y to conform with the usual notation for cubic curves.)

1.7.1 Show that the tangent to $y^2 = x^3 - 2$ at the "obvious" rational point $(3,5)$ is $y = \frac{27x}{10} - \frac{31}{10}$.

1.7.2 By substituting $y = \frac{27x}{10} - \frac{31}{10}$ in the equation of the curve, show that the tangent meets the curve where $100x^3 - 729x^2 + 1674x - 1161 = 0$.

1.7.3 By dividing $100x^3 - 729x^2 + 1674x - 1161$ twice by $x - 3$, or otherwise, show that the tangent meets the curve twice at $x = 3$ and once at $x = \frac{129}{100}$.

1.7.4 Hence find a rational point on $y^2 = x^3 - 2$ other than $(3, \pm 5)$.

There are in fact infinitely many rational points on the curve $y^2 = x^3 - 2$ (though this was not known until 1930; see Mordell (1969), Chapter 26), but we show later that its only integer points are $(3, \pm 5)$.

1.8   Gaussian integers

The Pythagorean equation appears in a new light if we use complex numbers to factorize the sum of two squares:

$$x^2 + y^2 = (x - yi)(x + yi) \quad \text{where } i = \sqrt{-1}.$$

Given that x and y are integers, the factors $x - yi$, $x + yi$ may be regarded as "complex integers". We denote the set of such "integers" by

$$\mathbb{Z}[i] = \{a + bi : a, b \in \mathbb{Z}\}$$

and call them the *Gaussian integers*, after Gauss, who was the first to realize that $\mathbb{Z}[i]$ has many properties in common with $\mathbb{Z}$.

For a start, it is clear that the sum, difference and product of numbers in $\mathbb{Z}[i]$ are also in $\mathbb{Z}[i]$, hence we can freely use $+$, $-$, and $\times$ and calculate by the same rules as in $\mathbb{Z}$. This already gives nice results about sums of squares and Pythagorean triples.

Two square identity. *A sum of two squares times a sum of two squares is a sum of two squares, namely*

$$(a_1^2 + b_1^2)(a_2^2 + b_2^2) = (a_1 a_2 - b_1 b_2)^2 + (a_1 b_2 + b_1 a_2)^2.$$

Proof. We factorize the sums of two squares as above, then recombine the two factors with negative signs, and the two factors with positive signs:

$$\begin{aligned}
(a_1^2 + b_1^2)(a_2^2 + b_2^2) &= (a_1 - b_1 i)(a_1 + b_1 i)(a_2 - b_2 i)(a_2 + b_2 i) \\
&= (a_1 - b_1 i)(a_2 - b_2 i)(a_1 + b_1 i)(a_2 + b_2 i) \\
&= [a_1 a_2 - b_1 b_2 - (a_1 b_2 + b_1 a_2)i] \times \\
&\qquad [a_1 a_2 - b_1 b_2 + (a_1 b_2 + b_1 a_2)i] \\
&= (a_1 a_2 - b_1 b_2)^2 + (a_1 b_2 + b_1 a_2)^2. \qquad \square
\end{aligned}$$

Corollary. *If the triples (a_1, b_1, c_1) and (a_2, b_2, c_2) are Pythagorean, then so is the triple $(a_1 a_2 - b_1 b_2, a_1 b_2 + b_1 a_2, c_1 c_2)$.*

Proof. If (a_1, b_1, c_1) and (a_2, b_2, c_2) are Pythagorean triples, then

$$a_1^2 + b_1^2 = c_1^2 \quad \text{and} \quad a_2^2 + b_2^2 = c_2^2.$$

It follows that

$$(c_1 c_2)^2 = c_1^2 c_2^2 = (a_1^2 + b_1^2)(a_2^2 + b_2^2)$$
$$= (a_1 a_2 - b_1 b_2)^2 + (a_1 b_2 + b_1 a_2)^2 \quad \text{by the identity above,}$$

and this says that $(a_1 a_2 - b_1 b_2, a_1 b_2 + b_1 a_2, c_1 c_2)$ is a Pythagorean triple.
$\square$

Of course, the two square identity can be proved without using $\sqrt{-1}$, by multiplying out both sides and comparing the results. And presumably it was first discovered this way, because it was known long before the introduction of complex numbers. Though first given explicitly by al-Khazin around 950 CE, it seems to have been known to Diophantus, and perhaps even to the Babylonians, because many of the triples implicit in Plimpton 322 can be obtained from smaller triples by the Corollary (see exercises).

However, the two square identity is more natural in the world $\mathbb{C}$ of complex numbers because it expresses one of their fundamental properties: namely, the *multiplicative property of their norm*. If $z = a + bi$ we define

$$\text{norm}(z) = |a + bi|^2 = a^2 + b^2,$$

and it follows from the two square identity that

$$\text{norm}(z_1)\text{norm}(z_2) = \text{norm}(z_1 z_2) \qquad (*)$$

because $z_1 = a_1 + b_1 i$ and $z_2 = a_2 + b_2 i$ imply

$$z_1 z_2 = a_1 a_2 - b_1 b_2 + (a_1 b_2 + b_1 a_2)i.$$

In algebra and complex analysis it is more common to state the multiplicative property (*) in terms of the *absolute value* $|z| = \sqrt{a^2 + b^2}$, namely

$$|z_1||z_2| = |z_1 z_2|. \qquad (**)$$

(*) and (**) are obviously equivalent, but the norm is the more useful concept in $\mathbb{Z}[i]$ because it is an ordinary integer, and this allows certain properties of $\mathbb{Z}[i]$ to be derived from properties of $\mathbb{Z}$.

So much for the elementary properties of Gaussian integers. $\mathbb{Z}[i]$ also has deeper properties in common with $\mathbb{Z}$, involving divisors and primes. These properties will be proved for $\mathbb{Z}$ in the next chapter, and for $\mathbb{Z}[i]$ in Chapter 6.

However, we can travel a little further in the right direction by following the dream that $\mathbb{Z}[i]$ holds the secrets of the Pythagorean equation

$$z^2 = x^2 + y^2 = (x - yi)(x + yi).$$

If the integers x and y have no common prime divisor, then it seems likely that $x - yi$ and $x + yi$ also have no common prime divisor, whatever "prime" means in $\mathbb{Z}[i]$. If so, then it would seem that the factors $x - yi$, $x + yi$ of the square z^2 are *themselves squares* in $\mathbb{Z}[i]$. In particular,

$$x - yi = (u - vi)^2 \quad \text{for some } u, v \in \mathbb{Z}.$$

But in that case

$$x - yi = (u^2 - v^2) - 2uvi$$

and, equating real and imaginary parts,

$$x = u^2 - v^2, \quad y = 2uv, \quad \text{and hence} \quad z = u^2 + v^2.$$

Thus we have arrived again at Euclid's formula for Pythagorean triples! (Or more precisely, the formula for *primitive* Pythagorean triples, from which all others are obtained as constant multiples. The primitive triples are those for which x, y, and z have no common prime divisor, and they result from u and v with no common prime divisor.)

The idea that factors of a square with no common prime divisor are themselves squares is essentially correct in $\mathbb{Z}[i]$, but to see why we must first understand why it is correct in $\mathbb{N}$. This will be explained in the next chapter.

Exercises

The rule in the Corollary for generating new Pythagorean triples from old gives some interesting results.

1.8.1 Find the Pythagorean triples generated from

- (4,3,5) and itself,
- (12,5,13) and itself,
- (15,8,17) and itself.

1.8.2 Do these results account for any of the entries in Plimpton 322?

1.8.3 Try to generate other entries in Plimpton 322 from smaller triples.

It is clear that we can generate infinitely many Pythagorean triples (x, y, z) but not clear (even from Euclid's formulas) whether there are any significant constraints on their members x, y, and z. For example, can we have x and y odd and z even? This question can be answered by considering remainders on division by 4.

1.8.4 Show that the square of an odd integer $2n + 1$ leaves remainder 1 on division by 4.

1.8.5 What is the remainder when an even square is divided by 4?

1.8.6 Deduce from Exercises 1.8.4 and 1.8.5 that the sum of odd squares is never a square.

1.9 Discussion

The discovery of Pythagorean triples, in which the sum $x^2 + y^2$ of two squares is itself a square, leads to a more general question: what values are taken by $x^2 + y^2$ as x and y run through $\mathbb{Z}$? The exercises above imply that $x^2 + y^2$ can *not* take a value of the form $4n + 3$ (why?), and the main problem in describing its possible values is to find the *primes* of the form $x^2 + y^2$.

Such questions were first studied by Fermat around 1640, sparked by his reading of Diophantus. He was able to answer them, and also the corresponding questions for $x^2 + 2y^2$ and $x^2 + 3y^2$. In the 18th century this led to study of the general *quadratic form* $ax^2 + bxy + cy^2$ by Euler, Lagrange, Legendre and Gauss. The endpoint of these investigations was the *Disquisitiones Arithmeticae* of Gauss (1801), a book of such depth and complexity that the best number theorists of the 19th century—Dirichlet, Kummer, Kronecker, and Dedekind—found that they had to rewrite it so that ordinary mortals could understand Gauss's results.

The reason that the *Disquisitiones* is so complex is that abstract algebra did not exist when Gauss wrote it. Without new algebraic concepts the deep structural properties of quadratic forms discovered by Gauss cannot be clearly expressed; they can barely be glimpsed by readers lacking the technical power of Gauss. It was precisely to comprehend Gauss's ideas and convey them to others that Kummer, Kronecker, and Dedekind introduced the concepts of rings, ideals, and abelian groups.

An intermediate step in the evolution of ring theory was the creation of *algebraic number theory*: a theory in which algebraic numbers such as $\sqrt{2}$ and i are used to illuminate the properties of natural numbers and integers. Around 1770, Euler and Lagrange had already used algebraic numbers to study certain Diophantine equations. For example, Euler successfully found all the integer solutions of $y^3 = x^2 + 2$ by factorizing the right-hand side into $(x + \sqrt{-2})(x - \sqrt{-2})$. He assumed that numbers of the form $a + b\sqrt{-2}$ "behave like" integers when a and b themselves are integers (see Section 7.1). The same assumption enables one to determine all primes of the form $x^2 + 2y^2$.

Such reasoning was rejected by Gauss in the *Disquisitiones*, since it was not sufficiently clear what it meant for algebraic numbers to "behave like" integers. In 1801 Gauss may already have known systems of algebraic numbers that did *not* behave like the integers. He therefore worked directly with quadratic forms and their integer coefficients, subduing them with his awesome skill in traditional algebra. However, Gauss (1832) took the first step towards an abstract theory of *algebraic integers* by proving that the Gaussian integers $\mathbb{Z}[i]$ do indeed "behave like" the ordinary integers $\mathbb{Z}$, specifically with respect to prime factorization. Among other things, this gives an elegant way to treat the quadratic form $x^2 + y^2$, as we see in Chapter 6.

The great achievement of Kummer and Dedekind was to tame the systems of algebraic numbers that do *not* behave like $\mathbb{Z}$, by adjoining new "numbers" to them. Kummer's mystery *ideal numbers*, and Dedekind's demystification of them in 1871, are among the most dramatic discoveries of mathematics. Ideal numbers also emerge naturally from the theory of quadratic forms, in particular from the form $x^2 + 5y^2$, so we follow the thread of quadratic forms throughout this book. Quadratic forms not only give the correct historical context for most of the concepts normally covered in ring theory but also provide the simplest and clearest examples.

2

The Euclidean algorithm

PREVIEW

Prime numbers may be regarded as the "building blocks" of the natural numbers because any natural number is a product of primes. (This explains, by the way, why 1 is *not* regarded as a prime—nothing is built from products of 1 except 1 itself). But even if primes are the building blocks, it is not easy to grasp them directly. There is no simple way to test whether a given natural number is prime, nor to find the smallest prime divisor of a given number.

Instead of studying the divisors of single numbers it is better to study the *common* divisors of pairs a, b. The ancient Euclidean algorithm is a remarkably efficient way to find the greatest common divisor (gcd) of given natural numbers a and b and it throws unexpected light on prime numbers and prime factorization.

It does so by representing $\gcd(a, b)$ as a *linear combination $ma + nb$*, where m and n are integers. This also leads to a clear understanding of the problem of solving linear equations in integers.

2.1   The gcd by subtraction

If natural numbers a and b have a common divisor d, then

$$a = a'd \quad \text{and} \quad b = b'd$$

for some natural numbers a' and b'. From this it follows that d divides $a - b$ because

$$a - b = a'd - b'd = (a' - b')d.$$

22

In other words, *a common divisor of a and b is also a divisor of a − b.*

Euclid used this fact to find the *greatest common divisor*, $\gcd(a,b)$, by "repeatedly subtracting the smaller number from the larger". More precisely, his algorithm goes as follows.

Suppose that $a > b$ and let

$$a_1 = a, \quad b_1 = b.$$

Then for each pair (a_i, b_i) we form the pair (a_{i+1}, b_{i+1}), where

$$a_{i+1} = \max(b_i, a_i - b_i), \quad b_{i+1} = \min(b_i, a_i - b_i).$$

Since this process produces smaller and smaller natural numbers, it must halt (by "descent"). We eventually get

$$a_k = b_k,$$

in which case we conclude that $\gcd(a,b) = a_k = b_k$.

The reason this algorithm works is that

$$\gcd(a_1, b_1) = \gcd(a_2, b_2) = \cdots = \gcd(a_k, b_k),$$

since any common divisor of the pair (a_1, b_1) is also a divisor of the pairs $(a_2, b_2), (a_3, b_3), \ldots, (a_k, b_k)$ produced by the successive subtractions.

Example. $a = 34$, $b = 19$

The algorithm gives the following pairs:

$$(a_1, b_1) = (34, 19)$$
$$(a_2, b_2) = (19, 34 - 19) = (19, 15)$$
$$(a_3, b_3) = (15, 19 - 15) = (15, 4)$$
$$(a_4, b_4) = (15 - 4, 4) = (11, 4)$$
$$(a_5, b_5) = (11 - 4, 4) = (7, 4)$$
$$(a_6, b_6) = (4, 7 - 4) = (4, 3)$$
$$(a_7, b_7) = (3, 4 - 3) = (3, 1)$$
$$(a_8, b_8) = (3 - 1, 1) = (2, 1)$$
$$(a_9, b_9) = (2 - 1, 1) = (1, 1)$$

and therefore $\gcd(34, 19) = \gcd(1, 1) = 1$.

Integer pairs a, b such that $\gcd(a,b) = 1$ are said to be *relatively prime*. Thus the Euclidean algorithm gives a simple means of deciding whether integers are relatively prime. In the next section we see that the algorithm (in a slightly modified form) is also highly efficient: it gives $\gcd(a,b)$ in a number of steps comparable with the total number of digits in a and b. It is harder to recognize whether a *single* integer n is prime: the obvious methods require a number of steps comparable with the *size* of n, which is exponentially larger—around 2^k if k is the number of binary digits of n.

Exercises

Starting with a pair of natural numbers and running the subtractive algorithm backwards—that is, repeatedly adding the two numbers most recently produced— gives what is called a *Lucas sequence*. The most famous of them is the *Fibonacci sequence* 1, 1, 2, 3, 5, 8, 13, ..., obtained by starting with the pair $(1,1)$.

2.1.1 Explain why the gcd of any two successive Fibonacci numbers is 1.

2.1.2 Consider the Lucas sequence that begins with 1, 3, 4, 7, 11, 18, 29, ... What is the gcd of any two successive terms?

The exponential difficulty of testing whether an integer n is prime can be seen in the case of the well-known method of trial division by integers $\leq \sqrt{n}$.

2.1.3 If n has a divisor $\neq 1, n$, explain why there must be such a divisor $\leq \sqrt{n}$.

2.1.4 If n has k digits in its binary numeral, show that there are at most $2^{k/2}$ numbers $\leq \sqrt{n}$. Can there be exactly $2^{k/2}$?

The fundamental fact about common divisors, that if d divides a and b then d divides $a \pm b$, throws light on primitive Pythagorean triples.

2.1.5 If $x = 2uv$ and $y = u^2 - v^2$, show that (x,y,z) is a primitive Pythagorean triple if and only of $\gcd(u,v) = 1$.

2.2   The gcd by division with remainder

Euclid's form of the gcd algorithm is usually speeded up by doing division with remainder instead of repeated subtraction. Given a pair (a_i, b_i) with $a_i > b_i$, the next pair is produced by the rule

$$a_{i+1} = b_i, \quad b_{i+1} = \text{remainder when } a_i \text{ is divided by } b_i$$

This is more efficient when a_i is many times as large as b_i, in which case many subtractions are replaced by one division.

However, the algorithm is essentially the same—division of natural numbers *is* just repeated subtraction—so it is still true that

$$\gcd(a_1, b_1) = \gcd(a_2, b_2) = \cdots$$

The only difference is that halting now occurs when b_k divides a_k, in which case we conclude that $\gcd(a, b) = \gcd(a_k, b_k) = b_k$.

Example. $a = 34$, $b = 19$ again.

The algorithm with division gives the following pairs:

$$(a_1, b_1) = (34, 19)$$
$$(a_2, b_2) = (19, 34 - 19) = (19, 15)$$
$$(a_3, b_3) = (15, 19 - 15) = (15, 4)$$
$$(a_4, b_4) = (4, 15 - 3 \times 4) = (4, 3)$$
$$(a_5, b_5) = (3, 4 - 3) = (3, 1)$$

Hence $\gcd(34, 19) = 1$ because 1 divides 3.

In this form of the algorithm it is easy to see that the number of divisions is comparable with the total number of digits in a and b. In fact, *if a and b are written in binary, then each division reduces the total number of digits by at least one.* If a has more digits than b this is clear: the new pair is b together with a remainder on division by b that has no more digits than b. If a and b have the same number of digits then, since both a and b necessarily begin with the digit 1, the remainder is simply $a - b$, and it has fewer digits than b.

The division form of the Euclidean algorithm is not only more efficient; it also has wider applicability. For example, in $\mathbb{Z}[i]$ we can divide 17 by $4 + i$ (exactly) and get the quotient $4 - i$, but it is meaningless to subtract $4 + i$ from 17 "$4 - i$ times". Thus division in $\mathbb{Z}[i]$ is not generally the result of repeated subtraction. Any Euclidean algorithm in $\mathbb{Z}[i]$ (and we see one in Section 6.4) necessarily uses division with remainder.

Exercises

The division form of the Euclidean algorithm on (a, b), where $a > b$, occurs when one finds what is called the *continued fraction* for a/b. The idea is that if $a = bq + r$, where $0 \leq r < b$, then

$$\frac{a}{b} = \frac{bq + r}{b} = q + \frac{r}{b} = q + \frac{1}{b/r},$$

and the process may then be repeated for the fraction b/r since $b > r$ by construction. However, the pair (b,r) is smaller than the initial pair (a,b), so this process terminates. The result is called the continued fraction for a/b.

2.2.1 Starting with

$$\frac{34}{19} = 1 + \frac{15}{19} = 1 + \frac{1}{19/15} = 1 + \frac{1}{1 + \frac{4}{15}},$$

show that the continued fraction for 34/19 is

$$1 + \cfrac{1}{1 + \cfrac{1}{3 + \cfrac{1}{1 + \frac{1}{3}}}}$$

2.2.2 Show similarly that

$$\frac{43}{30} = 1 + \cfrac{1}{2 + \cfrac{1}{3 + \frac{1}{4}}}.$$

2.2.3 Show in general that

$$\frac{a}{b} = q_1 + \cfrac{1}{q_2 + \cfrac{1}{q_3 + \cfrac{1}{\ddots}}}$$

where $q_1, q_2, q_3, \ldots$ are the successive quotients occurring when the division form of the Euclidean algorithm is applied to (a,b).

In the 18th century, Euler saw that the Euclidean algorithm could be implemented by continued fractions, and this became the favored way to describe it for a century or more. For example, Gauss ignores Euclid and refers exclusively to the "continued fraction" algorithm in his *Disquisitiones*. The Euclidean algorithm as we know it made a comeback with Dirichlet's *Vorlesungen über Zahlentheorie* (lectures on number theory) of 1863.

2.3   Linear representation of the gcd

Probably the most important consequence of the Euclidean algorithm is that

$$\gcd(a,b) = ma + nb \quad \text{for some integers } m \text{ and } n.$$

In fact it is true that *all the numbers a_i and b_i produced by the Euclidean algorithm are of the form $ma + nb$ for integers m and n*, and the b_i of course include $\gcd(a,b) = b_k$.

We prove this statement about a_i and b_i by the "ascent" form of induction. For a start, we certainly have

$$a_1 = 1 \times a + 0 \times b, \quad b_1 = 0 \times a + 1 \times b,$$

so the statement is true for $i = 1$. And if a_i and b_i are both of the form $ma + nb$ the same is true for their difference, hence for a_{i+1} and b_{i+1}. Thus all numbers produced from the pair (a,b) by the Euclidean algorithm are of the form $ma + nb$, as required. $\square$

This proof also suggests a way to find the m and n for a given a and b: run the Euclidean algorithm to find $\gcd(a,b)$ and keep track of the coefficients m and n for each number a_i and b_i that the algorithm produces.

A practical way to do this is shown in the following example, in which the numerical calculation of gcd on 34, 19 is run in parallel with a symbolic calculation on letters a, b. Each time we subtract some multiple of the second number from the first we do exactly the same operation on letters. Hence the final combination of letters equals the gcd.

Example. $\gcd(34, 19) = \gcd(a,b)$ in the form $ma + nb$. For efficiency, we use division with remainder, subtracting the appropriate multiple of the second number from the first to get the remainder at each step.

$$
\begin{aligned}
(34,19) &= (a,b) \\
\Rightarrow \quad (19,15) &= (b,a-b) \\
\Rightarrow \quad (15,4) &= (a-b, b-(a-b)) = (a-b, -a+2b) \\
\Rightarrow \quad (4,3) &= (-a+2b, a-b-3(-a+2b)) = (-a+2b, 4a-7b) \\
\Rightarrow \quad (3,1) &= (4a-7b, -a+2b-(4a-7b)) = (4a-7b, -5a+9b).
\end{aligned}
$$

From the last line we read off the gcd,

$$1 = -5a + 9b.$$

This checks, because

$$-5 \times 34 + 9 \times 19 = -170 + 171 = 1.$$

The Euclidean algorithm is extremely important in practice and theory. It is useful in practice because it is unusually fast—it gives the gcd of k-digit numbers in around k steps—much faster than any known algorithm for finding divisors of one k-digit number.

And the gcd is not only simpler in practice but also in theory. The basic theory of divisors and primes is based on the theory of the gcd, as we see in Section 2.4.

We often call on the Euclidean algorithm to find $\gcd(a,b)$ and to find integers m and n such that $\gcd(a,b) = ma + nb$. So make sure you get plenty of practice in using it right now!

Exercises

2.3.1 Find $\gcd(63, 13)$ by the Euclidean algorithm, and hence find m and n such that $63m + 13n = 1$.

2.3.2 Find m and n such that $55m + 34n = 1$

2.4   Primes and factorization

In Section 1.1 we used a descent argument to show that certain natural numbers have prime factors. A slight generalization of the argument shows:

Existence of prime factorization. *Each natural number n can be written as a product of primes,*

$$n = p_1 p_2 p_3 \cdots p_k.$$

Proof. If n itself is a prime there is nothing to do. If not, $n = ab$ for some smaller natural numbers a and b. If a or b is not prime we split it into smaller factors, and so on. Since natural numbers cannot decrease forever, we eventually get a factorization

$$n = p_1 p_2 p_3 \cdots p_k$$

in which no p_i is a product of smaller numbers. That is, each p_i is prime. $\square$

So much for the existence of prime factorization, which is important enough because it implies the existence of infinitely many primes, as in Section 1.1. Even more important is the *uniqueness* of prime factorization—no matter how we split n into smaller factors we always arrive at the same primes in the end.

Prime divisor property. *If a prime p divides the product of natural numbers a and b, then p divides a or p divides b.*

Proof. Suppose p does *not* divide a, so we need to show that p divides b.

Now if p does not divide a we have $\gcd(a, p) = 1$, since the only divisors of p are 1 and p. Therefore, by the result in Section 2.3,

$$1 = ma + np \quad \text{for some integers } m \text{ and } n.$$

Multiplying both sides of this equation by b we get

$$b = mab + npb.$$

Now look at the right-hand side: p divides ab by assumption, and p obviously divides pb. Thus p divides both terms on the right, and hence it divides their sum. That is, p divides b, as required.   □

Unique prime factorization.   *The prime factorization of each natural number is unique (up to the order of factors).*

Proof. Suppose on the contrary that some natural number has two different prime factorizations. Cancelling any primes common to both factorizations, we get equal products of primes,

$$p_1 p_2 p_3 \cdots p_k = q_1 q_2 q_3 \cdots q_l,$$

where no prime p_i equals any prime q_j. This leads to a contradiction as follows.

Since p_1 is a factor of the left-hand side, p_1 also divides the right-hand side. But then, by repeatedly using the prime divisor property, we get

$$p_1 \text{ divides } q_1 q_2 q_3 \cdots q_l$$
$$\Rightarrow p_1 \text{ divides } q_1 \quad \text{or} \quad p_1 \text{ divides } q_2 q_3 \cdots q_l$$
$$\Rightarrow p_1 \text{ divides } q_1 \quad \text{or} \quad p_1 \text{ divides } q_2 \quad \text{or} \quad p_1 \text{ divides } q_3 \cdots q_l$$
$$\vdots$$
$$\Rightarrow p_1 \text{ divides } q_1 \quad \text{or} \quad p_1 \text{ divides } q_2 \quad \ldots \quad \text{or} \quad p_1 \text{ divides } q_l$$
$$\Rightarrow p_1 = q_1 \quad \text{or} \quad p_1 = q_2 \quad \ldots \quad \text{or} \quad p_1 = q_l,$$

which contradicts the assumption that no p_i equals any q_j. Thus no natural number has two different prime factorizations.   □

Although the prime divisor property was proved by Euclid (around 300 BCE), unique prime factorization was mentioned for the first time by Gauss in 1801.

Exercises

Gauss proved unique prime factorization by a novel proof of the prime divisor property that goes as follows.

2.4.1 First show that a prime p cannot divide a product $a_1 b_1$ for natural numbers $a_1, b_1 < p$. Namely, suppose that p divides $a_1 b_1$, and show that p also divides $a_1 b_2$, where

$$b_2 = \text{remainder when } p \text{ is divided by } b_1,$$

which gives an infinite descent.

2.4.2 Now use Exercise 2.4.1 to deduce the prime divisor property by showing that if p divides ab, and p divides neither a nor b, then p divides an $a_1 b_1$ where $a_1, b_1 < p$.

2.5   Consequences of unique prime factorization

If $c = p_1^{m_1} p_2^{m_2} \cdots p_k^{m_k}$, where $p_1, p_2, \ldots, p_k$ are primes and $m_1, m_2, \ldots, m_k$ are natural numbers, then

$$c^2 = p_1^{2m_1} p_2^{2m_2} \cdots p_k^{2m_k}.$$

Thus *in the prime factorization of a square natural number each prime occurs to an even power.* And conversely, if

$$d = p_1^{2m_1} p_2^{2m_2} \cdots p_k^{2m_k}$$

then $d = c^2$. Thus in fact *a natural number is a square if and only if each prime in the prime factorization of d occurs to an even power.*

Now suppose that d is a square, and that $d = ab$, where a and b have no common prime divisor (or, as we said in Section 2.1, a and b are *relatively prime*). Then we have a prime factorization of the form

$$d = ab = p_1^{2m_1} p_2^{2m_2} \cdots p_k^{2m_k}.$$

Since a and b have no common prime divisor, each term $p_i^{2m_i}$ must be part of the prime factorization of one of a and b and completely absent from the other. In other words, in the prime factorizations of a and b each prime occurs to an even power, and hence *a and b are both squares* by the remark in the previous paragraph.

To sum up, we have the following proposition.

Relatively prime factors of a square. *If a and b are relatively prime natural numbers whose product is a square, then a and b are squares.* □

Using the fact that

$$c^3 = p_1^{3m_1} p_2^{3m_2} \cdots p_k^{3m_k},$$

there are similar proofs that *a natural number is a cube if and only each prime in its prime factorization occurs to a power that is a multiple of 3,* and that *if a and b are relatively prime natural numbers whose product is a cube, then a and b are cubes.*

Another important consequence of the prime factorization of a square is the existence of irrational square roots.

Irrational square roots. *If N is a nonsquare natural number, then $\sqrt{N}$ is irrational.*

Proof. Suppose that N is a natural number and that $\sqrt{N}$ is rational, that is,

$$\sqrt{N} = a/b \quad \text{for some natural numbers } a \text{ and } b.$$

We then have to show that N is a square. Squaring both sides, we get

$$N = a^2/b^2 = p_1^{2m_1} p_2^{2m_2} \cdots p_k^{2m_k}$$

for some primes $p_1, p_2, \ldots, p_k$. Each prime occurs to an even power, namely, twice its power in a minus twice its power in b. But then N is a square, as required, by the argument above (which also applies when some m_i are negative). □

Prime factorization, gcd, and lcm

Unique prime factorization implies that each prime divisor of a natural number n actually appears in the prime factorization of n. And any *common* prime divisor of a and b will appear in both their prime factorizations. Hence *the greatest common divisor of a and b is the product of the common primes in their prime factorizations.*

Examples.

$$666 = 2 \times 3^2 \times 37$$
$$1000 = 2^3 \times 5^3,$$

hence $\gcd(666, 1000) = 2$.

$$4444 = 2^2 \times 11 \times 101$$
$$9090 = 2 \times 3^2 \times 5 \times 101,$$

hence $\gcd(4444, 9090) = 2 \times 101 = 202$.

This method is quite effective for numbers that are small enough to factorize into primes. However, it is completely outclassed by the Euclidean algorithm for larger numbers. Also, it should be borne in mind that the factorization method is justified by unique prime factorization, which depends on the theory of the Euclidean algorithm.

Prime factorizations also give the *least common multiple* (lcm) of two natural numbers. Any common multiple of a and b must be a multiple of each prime power in a and b, hence *the least common multiple of a and b is the product of the maximum prime powers occurring in their prime factorizations*.

Examples. Using the factorizations of 666, 1000 and 4444, 9090 given above, we find

$$\operatorname{lcm}(666, 1000) = 2^3 \times 3^2 \times 5^3 \times 37 = 333000,$$

and
$$\operatorname{lcm}(4444, 9090) = 2^2 \times 3^2 \times 5 \times 11 \times 101 = 199980.$$

Exercises

As mentioned in Section 1.3, the concept of prime number is more complicated in $\mathbb{Z}$ than in $\mathbb{N}$, because the unit -1 can be part of a factorization. This complicates the situation with squares and cubes in $\mathbb{Z}$, but only slightly.

2.5.1 If a and b are relatively prime integers whose product is a square, show by means of an example that a and b are *not* necessarily squares. If they are not squares, what are they?

2.5.2 On the other hand, if a and b are relatively prime integers whose product is a cube, then a and b are cubes. Why?

The Euclidean algorithm shows immediately that $\gcd(2000, 2001) = 1$. Still it is interesting to actually see that 2000 and 2001 have no common prime divisor.

2.5.3 Find the prime factorizations of 2000 and 2001, thereby confirming that $\gcd(2000, 2001) = 1$.

It is also useful to have formulas for $\gcd(a,b)$ and $\mathrm{lcm}(a,b)$ in terms of the prime factorizations of a and b.

2.5.4 Suppose that $p_1, p_2, \ldots, p_k$ are all the primes that divide a or b, and that

$$a = p_1^{m_1} p_2^{m_2} \cdots p_k^{m_k},$$
$$b = p_1^{n_1} p_2^{n_2} \cdots p_k^{n_k}.$$

Deduce that

$$\gcd(a,b) = p_1^{\min(m_1,n_1)} p_2^{\min(m_2,n_2)} \cdots p_k^{\min(m_k,n_k)}$$
$$\mathrm{lcm}(a,b) = p_1^{\max(m_1,n_1)} p_2^{\max(m_2,n_2)} \cdots p_k^{\max(m_k,n_k)}.$$

2.5.5 Deduce from Exercise 2.5.4 that $\gcd(a,b)\mathrm{lcm}(a,b) = ab$.

Now that we know uniqueness of prime factorization, we can revisit Euclid's theorem about perfect numbers, mentioned in the exercises to Section 1.5.

2.5.6 If $2^p - 1 = q$ is prime, show that the proper divisors of $2^{p-1}q$ (those less than it) are $1, 2, 2^2, \ldots 2^{p-1}$ and $q, 2q, 2^2q, \ldots, 2^{p-2}q$.

2.5.7 Show that $1 + 2 + 2^2 + \cdots + 2^{p-2} = 2^{p-1} - 1$, and deduce that the sum of the proper divisors of $2^{p-1}q$ is $2^{p-1}q$. (That is, $2^{p-1}q$ is perfect.)

2.6   Linear Diophantine equations

The simplest nontrivial Diophantine equations are linear equations in two variables,

$$ax + by = c, \quad \text{where} \quad a, b, c \in \mathbb{Z}.$$

Such an equation may have infinitely many solutions or none. For example, the equation

$$6x + 15y = 0$$

has the infinitely many solutions $x = 15t$, $y = -6t$ as t runs through the integers. On the other hand, the equation

$$6x + 15y = 1$$

has no integer solutions. This is so because 3 divides $6x + 15y$ when x and y are integers (since 3 divides both 6 and 15) but 3 does not divide 1. This example shows that common divisors are involved in linear Diophantine equations, and exposes the key to their solution: the linear representation of the gcd found in Section 2.3.

Criterion for solvability of linear Diophantine equations. *When a, b, c are integers, the equation $ax + by = c$ has an integer solution if and only if $\gcd(a,b)$ divides c.*

Proof. Since $\gcd(a,b)$ divides a and b, it divides $ax + by$ for any integers x and y. Therefore, if $ax + by = c$, then $\gcd(a,b)$ divides c.

Conversely, we know from Section 2.3 that $\gcd(a,b) = am + bn$ for some integers m and n. Hence if $\gcd(a,b)$ divides c we have

$$c = \gcd(a,b)d = (am + bn)d = amd + bnd \quad \text{for some } d \in \mathbb{Z}.$$

But then $x = md$, $y = nd$ is a solution of $ax + by = c$.   □

This proof also shows how to find a solution $ax + by = c$ if one exists. Namely, express $\gcd(a,b)$ in the form $am + bn$, using the symbolic Euclidean algorithm to find m and n, then multiply m and n by the integer d such that $c = \gcd(a,b)d$.

If there is one solution $x = x_0$ and $y = y_0$, then there are infinitely many, because we can add to the pair (x_0, y_0) any of the infinitely many solutions of $ax + by = 0$.

General solution of $ax + by = c$. *The solution of $ax + by = c$ in $\mathbb{Z}$ is $x = x_0 + bt/\gcd(a,b)$, $y = y_0 - at/\gcd(a,b)$, where $x = x_0$, $y = y_0$ is any particular solution and t runs through $\mathbb{Z}$.*

Proof. Since $x = bt/\gcd(a,b)$, $y = -at/\gcd(a,b)$ is clearly an integer solution of $ax + by = 0$, adding it to any solution $x = x_0$, $y = y_0$ of $ax + by = c$ gives another solution of $ax + by = c$.

Conversely, if x, y is any solution of $ax + by = c$, then $x' = x - x_0$, $y' = y - y_0$ satisfies $ax' + by' = 0$. But any integer solution of $ax' + by' = 0$ is a solution of the equation

$$a'x' = -b'y'$$

whose coefficients are the relatively prime integers $a' = a/\gcd(a,b)$ and $b' = b/\gcd(a,b)$.

Since a' and b' have no common prime divisor, it follows from the unique prime factorization of both sides of the equation $a'x' = -b'y'$ that b' divides x'. That is

$$x' = b't \quad \text{for some integer } t, \text{ and hence} \quad y' = -a't.$$

Substituting the values of x', y', a', b' back in the equations above yields

$$x = x_0 + bt/\gcd(a,b), \quad y = y_0 - at/\gcd(a,b),$$

as claimed.   □

Exercises

The criterion for solvability can also be derived directly, by proving the following result without appeal to the Euclidean algorithm.

2.6.1 Show that $\{am + bn : m, n \in \mathbb{Z}\}$ consists of all integer multiples of $\gcd(a,b)$.

However, the Euclidean algorithm is invaluable for actually finding solutions to linear Diophantine equations.

2.6.2 Find an integer solution of $34x + 19y = 1$.

2.6.3 Also find an integer solution of $34x + 19y = 7$.

2.6.4 Is there an integer solution of $34x + 17y = 1$?

2.7   *The vector Euclidean algorithm

In Section 2.3 we used an extension of the Euclidean algorithm to compute the gcd of integers a and b in the form

$$\gcd(a,b) = ma + nb \quad \text{for some } m, n \in \mathbb{Z}.$$

The extension runs the ordinary algorithm ("subtracting the smaller number from the larger") and uses it to guide a *symbolic imitation* that performs the same operations on linear combinations of the letters a and b.

We now wish to analyze the symbolic part of the algorithm more closely in the case where a and b are relatively prime. To do so we replace each linear combination $m_i a + n_i b$ by the ordered pair, or *vector*, (m_i, n_i). To enable the ordinary algorithm to run as simply as possible we take $a > 0$ and $b < 0$ and keep the positive number in the first place and the negative in the second.

Then each step of the ordinary Euclidean algorithm is actually an *addition*: the number with the larger absolute value being replaced by its sum with the other number. The corresponding steps in the symbolic algorithm are *vector additions*, so we call the resulting process the *vector Euclidean algorithm*.

Example. Figure 2.1 shows the steps of the vector Euclidean algorithm on $(12, -5)$, with number pairs in the first column, symbolic pairs in the second column, and vector pairs in the third. The actual additions are shown only in the symbolic column.

Numbers	Symbolic pairs	Vector pairs
$(12,-5)$	(a,b)	$((1,0), (0,1))$
$(7,-5)$	$(a+b,b)$	$((1,1), (0,1))$
$(2,-5)$	$((a+b)+b,b) = (a+2b,b)$	$((1,2), (0,1))$
$(2,-3)$	$(a+2b,b+(a+2b)) = (a+2b,a+3b)$	$((1,2), (1,3))$
$(2,-1)$	$(a+2b,a+3b+(a+2b)) = (a+2b,2a+5b)$	$((1,2), (2,5))$
$(1,-1)$	$(a+2b+(2a+5b),2a+5b) = (3a+7b,2a+5b)$	$((3,7), (2,5))$

Figure 2.1: Outputs of Euclidean algorithms

From the bottom line we read off (as in Section 2.3) that

$$1 = 3a + 7b = 3 \times 12 - 7 \times 5$$

so $(m,n) = (3,7)$ is a natural number vector such that $12m - 5n = 1$.

It is also interesting to run the algorithm one step further (adding the number 1 to the number -1 in the first column to get 0), because 12 and 5 then reappear in the vector column.

| $(1,0)$ | $(3a+7b,2a+5b+(3a+7b)) = (3a+7b,5a+12b)$ | $((3,7), (5,12))$ |

Figure 2.2: Result of the extra step

This is not surprising because $0 = 5 \times 12 - 12 \times 5$, though conceivably we could have obtained a larger multiple of the vector $(5,12)$. What is interesting is how easily we arrive at the vector $(5,12)$: namely, *we started with the vectors* $\mathbf{i} = (1,0)$ *and* $\mathbf{j} = (0,1)$, *and took a series of steps in which a vector pair* $(\mathbf{v}_1, \mathbf{v}_2)$ *was replaced by either* $(\mathbf{v}_1 + \mathbf{v}_2, \mathbf{v}_2)$ *or* $(\mathbf{v}_1, \mathbf{v}_1 + \mathbf{v}_2)$.

We now generalize this example to show:

Relative primality in the vector Euclidean algorithm. *In running the vector Euclidean algorithm:*

1. *Every vector produced from* $(1,0)$ *and* $(0,1)$ *is a relatively prime pair of natural numbers.* (We call such a vector *primitive.*)

2. *Every relatively prime pair* (a,b) *of natural numbers can be produced (by starting the ordinary Euclidean algorithm on b and $-a$).*

Proof. 1. It is clear that any vector produced is a pair of natural numbers, because the first new pair is $(1,1)$ and further vector additions cannot decrease the members of the pair.

To see why each pair produced is relatively prime we prove a stronger property: *if* $((m_1,n_1),(m_2,n_2))$ *is the vector pair at any step, then*

$$m_1 n_2 - n_1 m_2 = 1.$$

This is true at the first step, when $(m_1,n_1) = (1,0)$ and $(m_2,n_2) = (0,1)$. And if it is true for the vector pair $((m_1,n_1),(m_2,n_2))$ then it is also true for the next pair $((m_1+m_2,n_1+n_2),(m_2,n_2))$ or $((m_1,n_1),(m_1+m_2,n_1+n_2))$. This is so because

$$(m_1+m_2)n_2 - (n_1+n_2)m_2 = m_1 n_2 - n_1 m_2 = 1$$

and

$$m_1(n_1+n_2) - n_1(m_1+m_2) = m_1 n_2 - n_1 m_2 = 1.$$

It follows that each vector (m_1,n_1) produced is a relatively prime pair, because any common divisor of m_1 and n_1 also divides $m_1 n_2 - n_1 m_2 = 1$. Similarly for each vector (m_2,n_2).

2. If a and b are relatively prime natural numbers then the vector Euclidean algorithm, guided by the ordinary Euclidean algorithm on b and $-a$, produces a vector (m,n) such that $mb - na = 0$, and m and n are relatively prime by part 1.

Since prime factorization is unique, $mb = na$ for relatively prime a, b and relatively prime m, n implies $m = a$ and $n = b$. Hence any relatively prime pair (a,b) can be produced by the vector Euclidean algorithm. $\square$

Exercises

The proof of relative primality in the vector Euclidean algorithm applies whether or not the guiding numbers b and $-a$ are relatively prime.

2.7.1 If b and $-a$ are *not* relatively prime, which vector (m,n) such that $mb = na$ is produced by the vector Euclidean algorithm?

As we saw in Section 2.6, the symbolic Euclidean algorithm is used when solving linear Diophantine equations. The above analysis of the vector algorithm directly shows its connection with certain equations. Suppose that we run the ordinary Euclidean algorithm on the numbers b and $-a$ until 1 and -1 are produced, and suppose that the corresponding vector pair is $((m_1,n_1),(m_2,n_2))$.

2.7.2 Show that $(x,y) = (m_1,n_1)$ is the *least* positive solution of $bx - ay = 1$ and that $(x,y) = (m_2,n_2)$ is the *least* positive solution of $bx - ay = -1$.

2.8   *The map of relatively prime pairs

The results of the previous section are presented graphically by Figure 2.3, which we call the *map of relatively prime pairs* or *primitive vectors*. It is a partition of the plane by an infinite tree into regions labelled by ordered integer pairs (a,b). The top two regions are labelled $(1,0)$ and $(0,1)$, and the other labels are generated by vector addition: if regions labelled $\mathbf{v}_1$ and $\mathbf{v}_2$ share an edge, then the region below the bottom end of the edge is labelled $\mathbf{v}_1 + \mathbf{v}_2$.

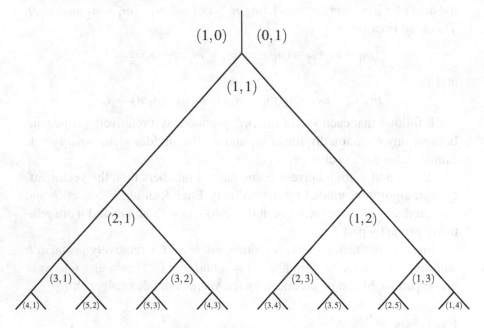

Figure 2.3: Regions labelled by relatively prime pairs

From the portion of the map shown in Figure 2.3, it appears that all labels are distinct and each of them $\neq (1,0)$ and $(0,1)$ is a relatively prime pair of natural numbers. This can be proved by relating the map to the vector Euclidean algorithm: the map is in fact a panoramic view of all outcomes of the algorithm, in the sense that *each sequence of vector pairs produced by a run of the algorithm occurs as the sequence of pairs of labels flanking the edges (to left and right) in a finite path down the tree.* This is so because both are governed by the vector addition rule.

Thus, the sequence $((1,0),(0,1)),((1,1),(0,1)),\dots,((3,7),(2,5))$ in the example of Section 2.7 is the sequence of left/right label pairs for the path shown in Figure 2.4.

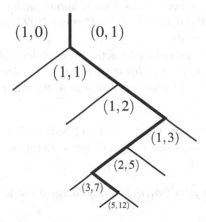

Figure 2.4: The branch leading to $(5,12)$

Conversely, any path down the tree, starting with the edge between $(1,0)$ and $(0,1)$ and ending at the top of region (a,b), is flanked by left/right pairs of labels that are precisely the pairs produced by the vector Euclidean algorithm with input numbers b and $-a$. This is so because the paths in a tree are unique, hence the path to the top vertex of region (a,b) must be the one corresponding to the vector Euclidean algorithm running on b and $-a$.

This correspondence between paths and runs of the vector Euclidean algorithm allows us to deduce the basic properties of the map from properties of the algorithm proved in the previous section.

1. *Each region of the map, except those labelled* $(1,0)$ *and* $(0,1)$, *is labelled by a relatively prime pair of natural numbers.* This follows from Property 1 of the vector Euclidean algorithm.

2. *Each relatively prime pair* (a,b) *of natural numbers occurs as a label.* This follows from Property 2 of the vector Euclidean algorithm.

3. *Each label occurs only once.* This is so because we reach the label (a,b) by running the ordinary Euclidean algorithm on b and $-a$, and the run determines a unique path in the tree.

Exercises

The map of relatively prime pairs has been discovered and rediscovered several times in the history of mathematics without ever becoming known well enough to acquire an official name. Perhaps its best known role is in representing rational numbers, since there is a one-to-one correspondence between positive rational numbers and *reduced fractions* a/b, which correspond in turn to relatively prime pairs (a,b) of natural numbers. This idea is known under the name of *Farey fractions*, and accounts of it may be found in Conway (1997), Rademacher (1983), and Hardy and Wright (1979).

The connection between reduced fractions and regions goes deeper than the obvious correspondence $a/b \leftrightarrow (a,b)$; it also *preserves order*. That is, *the ordering of fractions from large to small corresponds to the ordering of regions from left to right.*

2.8.1 Use the property $m_1 n_2 - n_1 m_2 = 1$ from Section 2.7 to show that if region (m_1,n_1) and (m_2,n_2) meet along an edge, with region (m_1,n_1) on the left, then $m_1/n_1 > m_2/n_2$.

2.8.2 Deduce that, if region (m_1,n_1) is *anywhere* to the left of (m_2,n_2), then $m_1/n_1 > m_2/n_2$.

2.8.3 Use Exercise 2.8.2 to give another proof that each label (a,b) occurs only once.

The tree structure of the Farey fractions is known as the Stern-Brocot tree. It is obtainable from our map by moving each label other than $(1,0)$ and $(0,1)$ to the vertex above it in Figure 2.3. More on the Stern-Brocot tree may be found in Graham *et al.* (1994).

We have taken our form of the map from Conway (1997), who uses it to give a very simple and graphic way of studying *quadratic forms*. For this purpose, the map has an advantage over the tree because it admits a natural extension to a map with regions labelled by all pairs of relatively prime *integers*. We take up this idea in Chapter 5.

2.9   Discussion

The results of this chapter are our answer to the question posed in Chapter 1 in connection with $\mathbb{Z}$ and $\mathbb{Z}[i]$: what does it mean to "behave like" the integers? Roughly speaking, the operations $+$ and $\times$ should make sense and have the ring properties, there should be primes, and there should be unique prime factorization (or equivalently, the prime divisor property).

The importance of unique prime factorization was first recognized by Gauss (1801) although, as mentioned in Section 2.4, the equivalent prime divisor property was known to Euclid. Another remarkable equivalent of unique prime factorization was discovered by Euler (1748a). It is his *product formula* for what is now called the *zeta function* $\zeta(s)$, defined by the following two expressions:

$$\sum_{n=1}^{\infty}\frac{1}{n^s} = \prod_{\text{primes } p}\left(\frac{1}{1-p^{-s}}\right). \qquad (*)$$

It is not obvious that these expressions are equal, and indeed their equality is equivalent to unique prime factorization! If we expand each factor on the right in a geometric series

$$\frac{1}{1-p^{-s}} = 1 + p^{-s} + p^{-2s} + p^{-3s} + \cdots,$$

then the product of all the factors will be the sum of 1 together with every possible term of the form

$$p_1^{-m_1 s} p_2^{-m_2 s} \cdots p_k^{-m_k s} = \frac{1}{\left(p_1^{m_1} p_2^{m_2} \cdots p_k^{m_k}\right)^s},$$

where $p_1, p_2, \ldots, p_k$ are distinct primes and $m_1, m_2, \ldots, m_k$ are natural numbers. The terms $p_1^{m_1} p_2^{m_2} \cdots p_k^{m_k}$ include each natural number n exactly once, just in case unique prime factorization holds, in which case we get the formula $(*)$.

What makes the product formula $(*)$ even more amazing is that it *also implies the infinitude of primes*, thus unifying the two most important theorems about primes. Euler's proof of infinitude uses the special value $s = 1$. If there are only finitely many primes, then the right-hand side of $(*)$ is finite for $s = 1$, whereas the left-hand side is $1 + \frac{1}{2} + \frac{1}{3} + \frac{1}{4} + \cdots$, which is well known to be infinite. Thus we have a contradiction, so there must be infinitely many primes.

The Euclidean algorithm was historically decisive for unique prime factorization, establishing this property for $\mathbb{Z}$, $\mathbb{Z}[i]$ and several other rings we meet later. Even before unique prime factorization was noticed, the algorithm was used by ancient Indian and Chinese mathematicians to solve linear Diophantine equations. Such equations arise in "calendar problems", where one has, say, a year of $365\frac{1}{4}$ days and a lunar cycle of $29\frac{1}{2}$ days and one wants to know such things as the next time that there will be a new moon on the first day of the year.

The modern history of the Euclidean algorithm begins with the discovery of Gauss (1832) that it also applies to $\mathbb{Z}[i]$. Dirichlet made the algorithm the basis of his *Vorlesungen* of 1863, using it to derive the basic results about $\mathbb{Z}$ in much the same way as we have here. The *Vorlesungen* went through four editions, evolving after Dirichlet's death through the editorial work of Dedekind, who began to enlarge it with Supplements from 1871 onwards. In the successive versions of Supplements X and XI Dedekind gradually freed number theory from dependence on the Euclidean algorithm by developing *ideal theory*, a development we take up in the last few chapters of this book.

3

Congruence arithmetic

PREVIEW

Many questions in arithmetic reduce to questions about remainders that can be answered in a systematic manner. For each integer $n > 1$ there is an arithmetic "mod n" that mirrors ordinary arithmetic but is *finite*, since it involves only the n remainders $0, 1, 2, \ldots, n-1$ occurring on division by n. Arithmetic mod n, or *congruence arithmetic*, is the subject of this chapter.

We motivate congruence arithmetic with some arithmetic folklore: the test for divisibility by 9 by "casting out nines". This is explained by the arithmetic of $+$, $-$, and $\times$ mod 9, and it leads naturally to $+$, $-$, and $\times$ mod n, and to the problem of division mod n. It turns out that division (by nonzero numbers) is possible mod n when n is prime, but not generally.

Division by a nonzero number a, mod n reduces to the problem of finding an *inverse* of a, mod n, that is, finding a b such that ab leaves remainder 1 on division by n. This turns out to be a simple spinoff of the procedure used in Chapter 2 to find integers m and n such that $ma + nb = \gcd(a,b)$, using the Euclidean algorithm.

Related to the subtleties of division are the classical theorems of Fermat, Euler, and Wilson, which are important throughout number theory and its applications. The most famous application, the RSA cryptosystem, is discussed in the next chapter, but the present chapter paves the way for it.

We also pave the way for studying *quadratic forms* $ax^2 + bxy + cy^2$ by using congruence arithmetic to show that certain values are impossible for the forms $x^2 + y^2$, $x^2 + 2y^2$, and $x^2 + 3y^2$.

3.1   Congruence mod n

Casting out nines

An old rule to test whether a natural number is divisible by 9 is to see whether the sum of its digits is divisible by 9. For example, 774 is divisible by 9 because

$$7 + 7 + 4 = 18,$$

which is divisible by 9.

This rule, called *casting out nines*, not only decides divisibility but in fact gives the remainder on division by 9. For example, if we add the digits of 476 we get

$$4 + 7 + 6 = 17,$$

which leaves remainder 8 on division by 9. This is also the remainder when 476 is divided by 9.

Now of course 476 does not stand for $4 + 7 + 6$ but for

$$4 \times 10^2 + 7 \times 10 + 6.$$

Yet somehow, as far as remainders are concerned, $4 + 7 + 6$ *behaves like* $4 \times 10^2 + 7 \times 10 + 6$.

To explain how this happens, we introduce the concept of congruence.

Definition. Integers a and b are said to be *congruent mod n*, written

$$a \equiv b \quad (\bmod\ n),$$

if they leave the same remainder on division by n. Equivalently, a is congruent to b, mod n, if n divides $a - b$.

We also say that a and b *belong to the same congruence class*, mod n.

Congruence mod 2 is the most familiar type of congruence in daily life, where we have words for numbers congruent to 0 (the *even* numbers), the numbers congruent to 1 (the *odd* numbers), and for numbers in the same congruence class (they have the same *parity*).

Congruence mod 2 is easy to recognize in decimal notation, as is congruence mod 5 and 10. We can tell immediately, for example, that 1244788 is even, 1244785 is divisible by 5, and 1244780 is divisible by 10. This is because numbers are congruent mod 2, 5, or 10 if their last digits are congruent mod 2, 5, or 10 respectively.

Likewise, we can tell whether numbers are congruent mod 4 by looking at their last two digits, and similar results apply to congruence mod other products of 2 and 5.

Congruence mod 9, the concept relevant to casting out nines, is not so easy to understand. For this we need *congruence arithmetic*.

Exercises

The rules given above for recognizing congruence mod 2, 5, and 10 are easy to explain and generalize.

3.1.1 Explain why the remainder of any natural number on division by 2 is the same as the remainder of its last digit.

3.1.2 Why does the same apply to division by 5 and 10, but not to division by 4?

3.1.3 Show that the remainder of n on division by 4 is the same as the remainder of the number given by the last two digits of n.

3.1.4 How many digits determine the remainder on division by 8; by 16?

3.2   Congruence classes and their arithmetic

The integers that leave remainder a on division by n form what is called the *congruence class* of a,

$$\{nk + a : k \in \mathbb{Z}\},$$

which we denote by the natural notation $n\mathbb{Z} + a$ (or just $n\mathbb{Z}$ when $a = 0$). For example

$$2\mathbb{Z} = \{\text{even numbers}\},$$
$$2\mathbb{Z} + 1 = \{\text{odd numbers}\}.$$

Each congruence class is a set of equally spaced points along the number line. For example, the classes $3\mathbb{Z}$, $3\mathbb{Z} + 1$ and $3\mathbb{Z} + 2$ look like the white, grey and black points respectively in Figure 3.1.

Figure 3.1: The congruence classes mod 3

Such pictures suggest that if we add any point in $n\mathbb{Z}+a$ to any point in $n\mathbb{Z}+b$ we get a point in $n\mathbb{Z}+(a+b)$. We can also see this algebraically: any point in $n\mathbb{Z}+a$ is of the form $nk+a$ and any point in $n\mathbb{Z}+b$ is of the form $nl+b$, so their sum $n(k+l)+(a+b)$ is in $n\mathbb{Z}+(a+b)$.

Therefore, it is meaningful to define the *sum* of congruence classes by

$$(n\mathbb{Z}+a)+(n\mathbb{Z}+b) = n\mathbb{Z}+(a+b),$$

since we land in the class of $a+b$ whichever elements we add from the class of a and the class of b. Similarly, it is meaningful to define the difference of congruence classes by

$$(n\mathbb{Z}+a)-(n\mathbb{Z}+b) = n\mathbb{Z}+(a-b).$$

Finally we have a *product* of congruence classes defined by

$$(n\mathbb{Z}+a)(n\mathbb{Z}+b) = n\mathbb{Z}+ab,$$

although it is not so obvious that any element of $n\mathbb{Z}+a$ times any element of $n\mathbb{Z}+b$ will be an element of $n\mathbb{Z}+ab$. To see why, take any member $nk+a$ from $n\mathbb{Z}+a$ and any member $nl+b$ from $n\mathbb{Z}+b$. Their product is

$$\begin{aligned}(nk+a)(nl+b) &= n^2kl+nkb+nla+ab\\ &= n(nkl+kb+la)+ab,\end{aligned}$$

which is indeed a member of $n\mathbb{Z}+ab$.

Another way to handle addition of congruence classes is by "addition of congruences". If we have the congruences

$$a_1 \equiv a_2 \pmod{n} \tag{1}$$

and

$$b_1 \equiv b_2 \pmod{n} \tag{2}$$

then (1) says a_1 and a_2 are in the same congruence class, call it $n\mathbb{Z}+a$, and (2) says b_1 and b_2 are in the same congruence class, call it $n\mathbb{Z}+b$. Then it follows that the sums a_1+b_1 and a_2+b_2 belong to the same congruence class $n\mathbb{Z}+(a+b)$, hence

$$a_1+b_1 \equiv a_2+b_2 \pmod{n} \tag{3}$$

Congruence (3) is the result of "adding" congruences (1) and (2).

Similarly, we can show by subtraction and multiplication of congruence classes that (1) and (2) imply

$$a_1 - b_1 \equiv a_2 - b_2 \pmod{n} \tag{4}$$

("subtraction of congruences") and

$$a_1 b_1 \equiv a_2 b_2 \pmod{n} \tag{5}$$

("multiplication of congruences").

Remark. The system of congruence classes mod n, under the operations of $+$ and $\times$, is denoted by $\mathbb{Z}/n\mathbb{Z}$. This agrees with the quotient notation for groups (see *Elements of Algebra*, Section 7.8), since $n\mathbb{Z}$ is a subgroup of $\mathbb{Z}$ and the congruence classes $n\mathbb{Z} + a$ are the cosets of $n\mathbb{Z}$ in $\mathbb{Z}$. However, in this book, $\mathbb{Z}/n\mathbb{Z}$ has the additional structure given by the $\times$ operation.

Casting out nines again

Using arithmetic mod 9 we can now explain the method of casting out nines introduced in Section 3.1.

First note that

$$10 \equiv 1 \pmod{9},$$

and therefore

$$10^2 \equiv 1^2 \equiv 1 \pmod{9},$$
$$10^3 \equiv 1^3 \equiv 1 \pmod{9},$$

and so on, by multiplication of congruences.

For any integer a_i it follows, by multiplication of congruences, that

$$a_i 10^i \equiv a_i \pmod{9},$$

and finally, by addition of congruences, that

$$a_k 10^k + \cdots + a_1 10 + a_0 \equiv a_k + \cdots + a_1 + a_0 \pmod{9}. \tag{*}$$

But if $a_0, a_1, \ldots, a_k$ are between 0 and 9 (that is, they are all decimal "digits"), then $a_k 10^k + \cdots + a_1 10 + a_0$ is the number whose decimal numeral is $a_k \cdots a_1 a_0$.

Thus (*) says that, on division by 9, $a_k \cdots a_1 a_0$ leaves the same remainder as the sum $a_k + \cdots + a_1 + a_0$, as required for casting out nines.

Exercises

There is an identical rule (which could be called "casting out threes") for testing
divisibility by 3, and a very similar rule for testing divisibility by 11.

3.2.1 Show that the above argument applies to show that

$$a_k 10^k + a_{k-1} 10^{k-1} + \cdots + a_1 10 + a_0 \equiv a_k + a_{k-1} + \cdots + a_1 + a_0 \quad (\text{mod } 3),$$

and hence that a number is divisible by 3 if and only if the sum of its digits
is divisible by 3.

3.2.2 Use $10 \equiv -1 \ (\text{mod } 11)$ to find what $10^2, 10^3, \ldots$ are congruent to, mod 11.

3.2.3 Deduce from Exercise 3.2.2, using multiplication and addition of congruences, that $a_k a_{k-1} \cdots a_1 a_0$ is divisible by 11 if and only if the "alternating
sum" of its digits, $(-1)^k a_k + \cdots + a_2 - a_1 + a_0$, is divisible by 11.

3.3   Inverses mod p

In $\mathbb{Z}$, the equation $ab = 1$ has only two solutions: $a, b = 1$ and $a, b = -1$.
Another way to put this is that 1 and -1 are the only integers with multiplicative inverses.

The situation is more interesting mod p, for prime p. In this case, *if*
$a \not\equiv 0 \ (\text{mod } p)$ *then there is a number b such that*

$$ab \equiv 1 \quad (\text{mod } p).$$

We say that *each $a \not\equiv 0$ (mod p) has a multiplicative inverse*, mod p.

Example. $p = 5$

1 has inverse 1,   2 has inverse 3,   3 has inverse 2,   4 has inverse 4.

The condition $a \not\equiv 0 \ (\text{mod } p)$ means that p does not divide a. Since p
is prime, it follows that $\gcd(a, p) = 1$. By Section 2.3, this implies that

$$ma + np = 1$$

for some $m, n \in \mathbb{Z}$. In other words,

$$ma \equiv 1 \quad (\text{mod } p),$$

so m is the required inverse of a, mod p.   □

Thus we can find the inverse m of a from the calculation (based on the
Euclidean algorithm) that finds the m and n such that $\gcd(a, b) = ma + nb$.
It follows that the computation of an inverse mod p is fast—it takes about
n steps for an n digit prime p.

Groups

The existence of inverses for all the nonzero congruence classes mod p implies that these congruence classes form a *group*, a concept briefly mentioned in Section 1.3 that we now review.

A group is a set G together with an operation on it, the *group operation*, with the *associative*, *identity*, and *inverse* properties. If the group operation is written as multiplication, then the identity element is written 1, the inverse of $g \in G$ is written g^{-1}, and the three properties are:

$$g_1(g_2 g_3) = (g_1 g_2)g_3 \qquad \text{(Associativity)}$$
$$g1 = 1g = g \qquad \text{(Identity property)}$$
$$gg^{-1} = g^{-1}g = 1 \qquad \text{(Inverse property)}$$

Now we can formally confirm that the nonzero congruence classes mod p form a group under multiplication. We call this group $(\mathbb{Z}/p\mathbb{Z})^{\times}$.

Group properties of $(\mathbb{Z}/p\mathbb{Z})^{\times}$. *For a prime p, the nonzero congruence classes mod p form a group under multiplication.*

Proof. First note that multiplication of congruence classes "inherits" associativity from the associativity of multiplication in $\mathbb{Z}$ as follows:

class of $a \times$ (class of $b \times$ class of c)
$= $ class of $a(bc)$ by definition
$= $ class of $(ab)c$ since $a(bc) = (ab)c$ by associativity in $\mathbb{Z}$
$= $ (class of $a \times$ class of b) $\times$ class of c by definition.

It follows that the product of nonzero congruence classes, mod p, is again a nonzero class. If $ab \equiv 0 \pmod{p}$ and we multiply both sides by the inverse c of b we get $(ab)c \equiv 0 \times c \equiv 0 \pmod{p}$ by multiplication of congruences. Hence the right-hand side 0 is congruent to the left-hand side

$(ab)c \equiv a(bc) \pmod{p}$ by associativity
$\equiv a(1) \pmod{p}$ since c is inverse to b
$\equiv a \pmod{p}$.

Thus the product is zero only when a factor is zero, hence the set of nonzero congruence classes is closed under multiplication, mod p.

We also have an identity element, namely the class of 1, and every element has an inverse by assumption. Thus $(\mathbb{Z}/p\mathbb{Z})^\times$ has all the defining properties of a group. $\qquad\qquad\qquad\qquad\qquad\qquad\qquad\qquad\qquad\qquad\qquad\quad\square$

$(\mathbb{Z}/p\mathbb{Z})^\times$ has the additional property that characterizes *abelian groups*:

$$g_1 g_2 = g_2 g_1 \qquad\qquad\qquad \text{(Commutativity)}$$

Most of the groups in this book are abelian, but the first theorem that we use—*Lagrange's theorem*—is easily proved in full generality. The proof is based on the concepts of *subgroup* and *cosets*.

A subset H of G that forms a group under the group operation in G is called a *subgroup of G*, and the (left) *cosets* of H in G are the sets of the form

$$gH = \{gh : h \in H\},$$

for all $g \in G$. Different $g_1, g_2 \in G$ do not necessarily produce different cosets $g_1 H$, $g_2 H$. For example, $h_0 H = H$ for any $h_0 \in H$, because each $h_0 h \in H$ when $h \in H$, and conversely each $h_1 \in H$ is of the form $h_0 h$ for some $h \in H$, namely $h = h_0^{-1} h_1$.

In fact, the proof we are about to give shows that the number of cosets gH for a subgroup H of a finite group G is precisely $|G|/|H|$, where $|G|$ and $|H|$ denote the "size" (that is, number of elements) of G and H respectively.

Lagrange's theorem. *If H is a subgroup of a finite group G, then $|H|$ divides $|G|$.*

Proof. First observe that each coset gH has the same size as H; the mapping from H to gH that sends h to gh can be reversed by multiplying on the left by g^{-1}. Thus all cosets have the same number of elements.

Second, we observe that any two cosets with a common element are identical. If $g \in g_1 H$ and $g \in g_2 H$ then

$$g = g_1 h_1 \text{ for some } h_1 \in H, \quad g = g_2 h_2 \text{ for some } h_2 \in H,$$

and therefore $g_1 h_1 = g_2 h_2$. Multiplying this on the right by h_1^{-1}, we find that $g_1 = g_2 h_2 h_1^{-1}$, and therefore

$$g_1 H = g_2 h_2 h_1^{-1} H = g_2 (h_2 h_1^{-1} H).$$

But $h_2 h_1^{-1} \in H$ and so, by the example preceding the proof, $h_2 h_1^{-1} H = H$. Hence $g_1 H = g_2 H$ as claimed.

These two observations together show that the $|G|$ elements of G fall into *disjoint* cosets gH of *equal size* $|H|$. Hence $|H|$ divides $|G|$. $\qquad\square$

Exercises

In the next section we use Lagrange's theorem to prove a famous theorem about congruence mod p. For readers not yet comfortable with group theory, the following exercises pave the way for a more direct proof using a minimum of information about inverses. Their content is a special case of the example preceding the proof of Lagrange's theorem—that multiplying a group by one of its elements reproduces the same set of elements.

Suppose that $a \not\equiv 0 \pmod{p}$, that is, a is not a multiple of p. Thus a has an inverse, mod p. Use it!

3.3.1 Show that $ia \equiv 0 \pmod{p} \Rightarrow i \equiv 0 \pmod{p}$.

3.3.2 Show that $ia \equiv ja \pmod{p} \Rightarrow i \equiv j \pmod{p}$.

3.3.3 Deduce from Exercises 3.3.1 and 3.3.2 that $a, 2a, 3a, \ldots, (p-1)a$ are distinct and $\not\equiv 0 \pmod{p}$, hence

$$\{a, 2a, 3a, \ldots, (p-1)a\} \equiv \{1, 2, 3, \ldots, p-1\} \pmod{p}.$$

3.3.4 Verify the result of Exercise 3.3.3 in the case $p = 7$, $a = 2$.

3.4   Fermat's little theorem

If we form powers $a, a^2, a^3, a^4, \ldots$ of any nonzero element a, mod p, then eventually there will be a repeated value, say

$$a^{m+n} \equiv a^m \pmod{p}.$$

Multiplying both sides by the inverse of a^m, mod p, then gives

$$a^n \equiv 1 \pmod{p}.$$

Thus in fact the series of powers always includes 1. For example, if we take $p = 5$ and $a = 2$ and compute $2, 2^2, 2^3, 2^4 \ldots$ mod 5 we find that $2^4 = 16 \equiv 1 \pmod{5}$. The sequence of powers repeats the same finite sequence $a, a^2, a^3, a^4, \ldots, a^{n-1}, 1$ forever and is therefore called *cyclic*.

From the group-theoretic point of view, the argument just given shows that the powers of a nonzero element, mod p, form a *subgroup* of the group $(\mathbb{Z}/p\mathbb{Z})^\times$. (Associativity and the identity element are obvious and the inverse of a^k is a^{n-k}.) Lagrange's theorem can then be applied, and it says how the size of the subgroup, and hence the least exponent n for which $a^n \equiv 1 \pmod{p}$, is related to p.

Fermat's little theorem. *If p is prime and a $\not\equiv 0$ (mod p), then*

$$a^{p-1} \equiv 1 \pmod{p}.$$

Proof. $(\mathbb{Z}/p\mathbb{Z})^{\times}$ has $p-1$ members, the classes of $1, 2, 3, \ldots, p-1$, so the size of any subgroup of $(\mathbb{Z}/p\mathbb{Z})^{\times}$ divides $p-1$ by Lagrange's theorem.

In particular, if $a \not\equiv 0$ (mod p) and $n > 1$ is the least exponent for which $a^n \equiv 1$ (mod p), then the powers of the class of a form a subgroup with n members, and hence n divides $p-1$.

But if

$$a^n \equiv 1 \pmod{p}$$

and n divides $p-1$ (say, $p-1 = mn$) then

$$a^{p-1} \equiv a^{mn} \equiv (a^n)^m \equiv 1^m \equiv 1 \pmod{p} \qquad \square$$

Application: a formula for the inverse mod p

It follows from Fermat's little theorem that, for any $a \not\equiv 0$ (mod p),

$$a^{p-2} \cdot a \equiv 1 \pmod{p},$$

hence a^{p-2} is the inverse of a, mod p. This is not only an explicit formula for the inverse mod p, it also implies an efficient method to compute it, competitive with the Euclidean algorithm method of the previous section.

We know from Section 1.5 that a^{p-2} can be computed in about $\log p$ multiplications, and here the numbers to be multiplied are $\leq p$, since we are working mod p. Compare this with finding the inverse of a by the method of Section 3.3: using the Euclidean algorithm to express $1 = \gcd(a, p)$ in the form $ma + np$, which gives m as the inverse of a, mod p. This involves about $\log p$ divisions with remainder (plus some other, less time-consuming, arithmetic), again on numbers $\leq p$. Since division takes about the same time as multiplication, the two methods are of similar speed.

Primitive roots

The *minimum* positive integer n such that $a^n \equiv 1$ (mod p) is called the *order* of a in $(\mathbb{Z}/p\mathbb{Z})^{\times}$. The proof tells us that the order of any nonzero a, mod p, is a divisor of $p-1$. There is always an a of order exactly $p-1$, called a *primitive root* for p. Its existence was conjectured by Euler and first proved by Gauss (1801). Primitive roots do not play an important role in this book, though sometimes they throw light on results provable by other means. Thus their properties and proof of existence are not essential reading, but they are in the starred sections at the end of this chapter.

Exercises

We now complete the proof of Fermat's little theorem begun in the previous exercise set.

3.4.1 Deduce from Exercise 3.3.3 that

$$a^{p-1} \times 1 \times 2 \times 3 \times \cdots \times (p-1) \equiv 1 \times 2 \times 3 \times \cdots \times (p-1) \quad (\bmod\ p).$$

3.4.2 Exercise 3.4.1 implies that $a^{p-1} \equiv 1 \pmod{p}$. Why?

Now for a few simple exercises on primitive roots.

3.4.3 Show that 2 is a primitive root for 5 but not for 7.

3.4.4 Find a primitive root for 7.

3.4.5 Given the existence of primitive root for p show that every divisor of $p-1$ occurs as the order of some element of $(\mathbb{Z}/p\mathbb{Z})^\times$.

3.5   Congruence theorems of Wilson and Lagrange

Another useful application of inverses mod p is the following theorem, which actually evaluates the product $(p-1)! = 1 \times 2 \times 3 \times \cdots \times (p-1)$ used in some proofs of Fermat's little theorem. It will be useful to know the value of $(p-1)! \bmod p$ in Section 9.8, when we come to the law of quadratic reciprocity. The theorem is credited to Wilson (and it may in fact have been discovered by Ibn al-Haytham in the 10th century), but the first known proof is due to Lagrange.

Wilson's theorem. *If p is prime then* $(p-1)! \equiv -1 \pmod{p}$.

Proof. In this congruence the factors $1, 2, 3, \ldots, p-1$ all have inverses mod p, hence each is cancelled by its own inverse except the factors that are inverse to themselves.

Such self-inverse factors x are 1 and $p-1 \equiv -1 \pmod{p}$, and no others, because if $x^2 \equiv 1 \pmod{p}$ we have

$$x^2 - 1 \equiv (x-1)(x+1) \equiv 0 \quad (\bmod\ p).$$

In other words, p divides $(x-1)(x+1)$. But then p divides $x-1$ or p divides $x+1$ by the prime divisor property, hence

$$x \equiv 1 \pmod{p} \quad \text{or} \quad x \equiv -1 \pmod{p}, \quad \text{as claimed.}$$

Thus the product $(p-1)!$ is $\equiv -1 \pmod{p}$, as required.   □

The fact that the congruence $x^2 - 1 \equiv 0 \pmod{p}$ has at most two solutions has an important generalization due to Lagrange.

Lagrange's polynomial congruence theorem. *If $P(x)$ is a polynomial of degree n with integer coefficients, and p is prime, then the congruence*

$$P(x) \equiv 0 \pmod{p}$$

has at most n incongruent solutions, mod p.

Proof. If there is no solution, we are done. Otherwise, suppose $P(r) \equiv 0 \pmod{p}$, where

$$P(x) = a_n x^n + a_{n-1} x^{n-1} + \cdots + a_1 x + a_0 \quad \text{and} \quad a_n, a_{n-1}, \ldots, a_1, a_0 \in \mathbb{Z}.$$

This implies

$$P(x) \equiv P(x) - P(r) \pmod{p}$$
$$\equiv a_n(x^n - r^n) + a_{n-1}(x^{n-1} - r^{n-1}) + \cdots + a_1(x - r) \pmod{p}$$
$$\equiv (x - r)Q(x) \pmod{p} \tag{*}$$

where $Q(x)$ is the polynomial of degree $n-1$ that remains when $x-r$ has been extracted from each of $x^n - r^n, x^{n-1} - r^{n-1}, \ldots, x-r$ using the identity

$$x^k - r^k = (x - r)(x^{k-1} + x^{k-2} r + \cdots + x r^{k-2} + r^{k-1}).$$

It follows from (*) and the prime divisor property that the congruence $P(x) \equiv 0 \pmod{p}$ implies

$$x - r \equiv 0 \pmod{p} \quad \text{or} \quad Q(x) \equiv 0 \pmod{p}.$$

Since $Q(x)$ has degree $n-1$, we can assume inductively that the congruence $Q(x) \equiv 0 \pmod{p}$ has at most $n-1$ incongruent solutions. Then

$$P(x) \equiv (x - r)Q(x) \equiv 0 \pmod{p}$$

has at most n incongruent solutions (namely $x = r$ and the solutions of $Q(x) \equiv 0 \pmod{p}$), as required.   □

Two important uses of this theorem are to prove the existence of primitive roots for p (Section 3.9), and to prove Euler's criterion for squares mod p (Section 9.3).

Exercises

Wilson's theorem actually gives a criterion for a natural number n to be prime.

3.5.1 If n is not prime, show that n divides $(n-1)!$, that is, $(n-1)! \equiv 0 \pmod{n}$.

3.5.2 Deduce from Exercise 3.5.1 that n is prime $\Leftrightarrow (n-1)! \equiv -1 \pmod{n}$.

3.5.3 Check that this criterion works when $n = 7$.

Unfortunately, the criterion has no practical value when n is large (say, 100 digits) because in this case we have no feasible way to compute $(n-1)! \bmod n$.

3.6   Inverses mod k

It is *not* always true that an $a \not\equiv 0 \pmod{k}$ has an inverse mod k.
    For example, $2 \not\equiv 0 \pmod{4}$ but

$$2 \times 2 = 4 \equiv 0 \pmod{4}.$$

Thus 2 has no inverse, for if it did we could multiply both sides of

$$2 \times 2 \equiv 0 \pmod{4}$$

by the inverse of 2 and get the false result $2 \equiv 0 \pmod{4}$.

Criterion for existence of an inverse, mod k. *An integer a has an inverse mod k if and only if $\gcd(a,k) = 1$.*

Proof. If $\gcd(a,k) = 1$ then, by Section 2.3,

$$\gcd(a,k) = 1 = ma + nk \quad \text{for some } m,n \in \mathbb{Z}.$$

This says that

$$ma \equiv 1 \pmod{k},$$

so m is an inverse of a, mod k.
    Conversely, if m is an inverse of a, mod k, then

$$ma \equiv 1 \pmod{k}.$$

Hence

$$ma + nk = 1 \quad \text{for some } m,n \in \mathbb{Z}.$$

This implies $\gcd(a,k) = 1$, because any common divisor of a and k also divides $ma + nk$, which equals 1. $\qquad\square$

If a_1 and a_2 have inverses m_1 and m_2 mod k, then a_1a_2 has inverse m_1m_2. It follows that the elements with inverses mod k form a set closed under multiplication and hence a group, which is called $(\mathbb{Z}/k\mathbb{Z})^\times$. (The group properties may be checked as they were for $(\mathbb{Z}/p\mathbb{Z})^\times$ in Section 3.3.)

Example. $(\mathbb{Z}/8\mathbb{Z})^\times$

1 has inverse 1, 3 has inverse 3, 5 has inverse $5\equiv -3$, 7 has inverse $7\equiv -1$, and it can be checked that these are the only invertible elements. Thus $(\mathbb{Z}/8\mathbb{Z})^\times$ is an abelian group with four elements. It is *not* cyclic, because each of its elements has order ≤ 2.

The size of $(\mathbb{Z}/k\mathbb{Z})^\times$, that is, the number of elements a among

$$1,2,3,\ldots,k-1$$

such that $\gcd(a,k) = 1$, is denoted by $\varphi(k)$ and is called the *Euler phi function*. For example, $\varphi(8) = 4$ because the four elements 1, 3, 5, 7 are the only natural numbers $a < 8$ for which $\gcd(a,8) = 1$.

Certain properties of φ are known, for example

- $\varphi(p^i) = p^{i-1}(p-1)$ for p prime,

- $\varphi(mn) = \varphi(m)\varphi(n)$ if $\gcd(m,n) = 1$.

These make it easy to compute $\varphi(k)$ if the prime factorization of k is known, but otherwise it is difficult.

If we apply Lagrange's theorem to an element a of $(\mathbb{Z}/k\mathbb{Z})^\times$, exactly as we did to an element a of $(\mathbb{Z}/p\mathbb{Z})^\times$ in Section 3.4, then we obtain the following.

Euler's theorem. *If a is invertible mod k then*

$$a^{\varphi(k)} \equiv 1 \pmod{k}.$$

Proof. We use the same argument as for Fermat's little theorem, except that now we use the fact that the size of the group $(\mathbb{Z}/k\mathbb{Z})^\times$ is $\varphi(k)$. $\square$

Like Fermat's little theorem does for $k = p$, Euler's theorem gives a formula for the inverse of a, mod k, namely $a^{\varphi(k)-1}$. The formula for general k is not quite so explicit because it involves the φ function. This blocks the computation of the inverse by exponentiation mod k because there is no efficient way known to compute $\varphi(k)$. In fact, the difficulty of computing $\varphi(k)$ is important for the security of the famous RSA cryptosystem studied in the next chapter.

Exercises

The formula $\varphi(p^i) = p^{i-1}(p-1)$ (and its special case when $i = 1$) may be shown as follows.

3.6.1 Explain why $\varphi(p) = p - 1$ when p is prime.

3.6.2 Show that there are p^{i-1} multiples of p among the numbers $1, 2, 3, \ldots, p^i$.

3.6.3 Deduce that $\varphi(p^i) = p^{i-1}(p-1)$ when p is prime.

The formula $\varphi(mn) = \varphi(m)\varphi(n)$ when $\gcd(m, n) = 1$ is proved in Section 9.7. For the time being we consider just a simple case.

3.6.4 Verify that $\varphi(15) = \varphi(3)\varphi(5)$.

3.7 Quadratic Diophantine equations

The behavior of quadratic Diophantine equations is much more complex than that of the linear Diophantine equations discussed in the last chapter. However, congruences are a good tool for showing that certain equations do *not* have solutions of a certain form.

Example 1. $x^2 + y^2 = p$ has no solution for p of the form $4n + 3$.

This statement is equivalent to $x^2 + y^2 \not\equiv 3 \pmod 4$, which we can prove by trying the finitely many values of x and $y \pmod 4$. These are $x, y \equiv 0$, 1, 2, -1, for which we have $x^2, y^2 \equiv 0, 1$.

It follows that $x^2 + y^2 \equiv 0, 1, 2 \pmod 4$, and so $x^2 + y^2 \not\equiv 3 \pmod 4$, as claimed. □

Example 2. $x^2 + 2y^2 = p$ has no solution for p of the form $8n + 5, 8n + 7$.

This statement is equivalent to $x^2 + 2y^2 \not\equiv 5, 7 \pmod 8$, which we can prove by trying the finitely many values of x and $y \pmod 8$. These are $x, y \equiv 0, 1, 2, 3, 4, -3, -2, -1$, for which we have $x^2, y^2 \equiv 0, 1, 4$.

It follows that $x^2 + 2y^2 \equiv 0, 1, 2, 3, 4, 6 \pmod 8$, so $x^2 + 2y^2 \not\equiv 5, 7$, $\pmod 8$, as claimed. □

Example 3. $x^2 + 3y^2 = p$ has no solution for p of the form $3n + 2$.

This statement is equivalent to $x^2 + 3y^2 \not\equiv 2 \pmod 3$, which we can prove by trying the finitely many values of x and $y \pmod 3$. These are $x, y \equiv 0, 1, -1$, for which we have $x^2, y^2 \equiv 0, 1$.

It follows that $x^2 + 3y^2 \equiv x^2 \equiv 0, 1 \pmod 3$, so $x^2 + 3y^2 \not\equiv 2 \pmod 3$, as claimed. □

These three results were first claimed by Fermat, though he credited them to his secret weapon, the "method of descent", apparently overlooking the easy congruence proofs. Descent is much heavier artillery (we use it in Section 7.7 to prove that $x^3 + y^3 \neq z^3$ for natural numbers x, y and z) and Fermat used it appropriately to prove more difficult *complements* of the results just mentioned. For example, while $x^2 + y^2$ never takes a prime value of the form $4n + 3$ (by the argument above), it takes *every* prime value of the form $4n + 1$.

Fermat became interested in *primes* of the form $x^2 + y^2$, $x^2 + 2y^2$ and $x^2 + 3y^2$ (which is why we denoted the right-hand side of the equations above by p) after reading a remark of Diophantus (*Arithmetica*, Book III, Problem 19):

> 65 is naturally divided into two squares in two ways, namely into $7^2 + 4^2$ and $8^2 + 1^2$, which is due to the fact that 65 is the product of 13 and 5, each of which is the sum of two squares.

Evidently Diophantus was aware of the formula

$$(a_1^2 + b_1^2)(a_2^2 + b_2^2) = (a_1 a_2 \pm b_1 b_2)^2 + (b_1 a_2 \mp a_1 b_2)^2,$$

which shows that the product of sums of two squares is itself the sum of two squares (in two different ways, corresponding to the choice of sign on the right-hand side).

Fermat saw what this implies: knowing which natural numbers are sums of two squares depends on knowing which primes are sums of two squares. The easy congruence argument in Example 1 shows that primes of the form $4n + 3$ are *not* sums of two squares; the hard part is to show that all primes of the form $4n + 1$ *are* sums of two squares. The theorem became something of a showcase for new methods in number theory, with Lagrange, Gauss and others using it to show off their innovations. In Chapter 6 we give a proof using the Gaussian integers, due to Dedekind.

It is also true that the primes of the form $x^2 + 2y^2$ are precisely those of the forms $8n + 1$ and $8n + 3$ *not* ruled out by the congruence arguments above (of course, numbers of the form $8n + 2$, $8n + 4$, or $8n + 6$ are not primes because they are divisible by 2). And likewise, the primes of the form $x^2 + 3y^2$ are those of the form $3n + 1$. We prove these results later by combining results from Chapter 7 and Chapter 9.

Exercises

It is entertaining to test Fermat's two square theorem on the first few primes of the form $4n + 1$ and to investigate his corresponding theorems on primes of the form $8n + 1$, $8n + 3$, and $3n + 1$.

3.7.1 Write down the first 10 primes of the form $4n + 1$ and check that each of them is a sum of two squares. (The first is $5 = 2^2 + 1^2$.)

3.7.2 Is any of these the sum of squares in two different ways?

3.7.3 Write down the first 10 primes of the form $8n + 1$ or $8n + 3$ and check that each of them is of the form $x^2 + 2y^2$. (And see whether any of them is of this form in two different ways.)

3.7.4 Write down the first 10 primes of the form $3n + 1$ and check that each of them is of the form $x^2 + 3y^2$. (And see whether any of them is of this form in two different ways.)

3.8 *Primitive roots

An interesting and puzzling phenomenon in elementary arithmetic is the *period in the decimal expansion of* $1/n$. For example, we know that

$$1/3 = 0.3333 \cdots$$
$$1/7 = 0.142857\,142857 \cdots$$
$$1/13 = 0.076923\,076923 \cdots$$

We say that the decimal of $1/3$ has *period length* 1 because the 1-digit pattern 3 repeats; $1/7$ has *period length* 6 because the 6-digit pattern 142857 repeats; and $1/13$ likewise has period length 6 because the 6-digit pattern 076923 repeats. It is clear from the ordinary school division process that a repetition must eventually occur, so periodicity in the decimal expansion of $1/n$ is not surprising. But why is the maximum period length $n - 1$, and under what circumstances does this occur?

The main part of the answer is that period length $n - 1$ occurs when 10 has *order* $n - 1$ in the group $(\mathbb{Z}/n\mathbb{Z})^\times$, that is, when 10^{n-1} is the least positive power of 10 that is $\equiv 1 \pmod{n}$. We also express this condition by saying that 10 is a *primitive root* for n. A closer study of $(\mathbb{Z}/n\mathbb{Z})^\times$, using Euler's theorem, then shows why $n - 1$ is the maximum possible period length.

Example. $1/7 = 0.142857\,142857 \cdots$

If we multiply this equation by 10, 10^2, 10^3, ... then we obtain

$$10/7 = 1.42857\,142857\cdots$$
$$10^2/7 = 14.2857\,142857\cdots$$

$$\vdots$$

$$10^6/7 = 142857.142857\cdots = 142857 + 1/7$$

Thus 10^6, like $10^0 = 1$, leaves remainder 1 on division by 7, and it is the first among the positive powers of 10 with this property (because $10^i/7$ has a different decimal part for $i = 1,2,3,4,5$). This is precisely what it means for 10 to be a primitive root for 7.

A generalization of this argument gives the following.

Criterion for maximal period length. *The decimal expansion of $1/n$ is periodic of length $n-1$ precisely when 10 is a primitive root for n. Also, $n-1$ is the maximum possible period length, occurring only when n is prime.*

Proof. Suppose that $1/n$ has a periodic decimal expansion with period length $n-1$,
$$1/n = 0.a_1 a_2 \cdots a_{n-1} a_1 a_2 \cdots a_{n-1} \cdots$$
If we multiply this equation by 10, 10^2, 10^3, ... then we get

$$10/n = a_1.a_2 a_3 \cdots a_{n-1} a_1 a_2 \cdots a_{n-1} \cdots$$
$$10^2/n = a_1 a_2.a_3 \cdots a_{n-1} a_1 a_2 \cdots a_{n-1} \cdots$$

$$\vdots$$

$$10^{n-1}/n = a_1 a_2 \cdots a_{n-1}.a_1 a_2 \cdots a_{n-1} \cdots = a_1 a_2 \cdots a_{n-1} + 1/n.$$

Thus 10^{n-1} is the first among the powers 10, 10^2, 10^3, ... that leaves remainder 1 on division by n (because $10^i/n$ has different decimal part when $i < n-1$). That is, 10 has order $n-1$ and hence is a primitive root for n.

Conversely, if 10 has order $n-1$ in $(\mathbb{Z}/n\mathbb{Z})^\times$ then n is prime. This follows from the proof of Euler's theorem, which shows that the order of any element of $(\mathbb{Z}/n\mathbb{Z})^\times$ is at most $\varphi(n)$. It is clear from the definition of the phi function that $\varphi(n) \le n-1$, and that equality holds only if n is prime.

It remains to show that the decimal of $1/n$ is periodic of length $n-1$ when 10 has order $n-1$. This follows by considering $1/n$, $10/n$, $10^2/n$,...,

$10^{n-1}/n$ again. The assumption that 10 has order $n-1$ implies that 10^{n-1} leaves remainder 1 on division by n, so we have

$$10^{n-1}/n = a_1a_2\cdots a_{n-1} + 1/n, \qquad (*)$$

where $a_1a_2\cdots a_{n-1}$ is the decimal numeral consisting of the first $n-1$ digits of $1/n$. Dividing both sides by 10^{n-1}, it follows that

$$1/n = 0.a_1a_2\cdots a_{n-1}\,a_1a_2\cdots a_{n-1}\cdots, \qquad (**)$$

though it is not clear what appears after the first $2(n-1)$ digits on the right. Repeatedly substituting $(**)$ back in $(*)$ and dividing by 10^{n-1} we find that the sequence $a_1a_2\cdots a_{n-1}$ keeps repeating in the decimal of $1/n$. This sequence defines the period length $n-1$ of $1/n$, because if there were a period of shorter length k we could conclude as above that 10 has order $k < n-1$, contrary to assumption.   □

In the exercises below you are asked to find a prime $p > 7$ for which 10 is a primitive root for p, and hence $1/p$ has period length $p-1$. In 1801 Gauss conjectured that the maximum period length $p-1$ occurs for infinitely many primes p but it is still not known whether this is true. In fact, it is not known whether *any* specific number, say 2 or 3, is a primitive root for infinitely many primes p. However, it is known that each prime p has a primitive root. We give a proof of this theorem in the next section.

Exercises

The decimal expansion of $1/n$ need not be periodic when n is not prime, for example $1/6 = 0.1666666\ldots$. The latter decimal expansion is called *ultimately periodic* because it is periodic beyond a certain digit (in this case, beyond the first digit).

3.8.1 Compute the decimal expansions of 1/12 and 1/14 by hand and verify that they are ultimately periodic.

3.8.2 Explain in general why $1/n$ has an ultimately periodic decimal expansion.

The relationship between decimal expansions and powers of 10 allows us to use properties of the decimal expansion of $1/n$ to predict properties of powers of 10, mod n, and vice versa.

3.8.3 Show, without using the decimal for 1/13, that 10 has order 6 in $(\mathbb{Z}/13\mathbb{Z})^\times$.

3.8.4 Which is the first prime $p > 7$ for which 10 is a primitive root for p? Verify that, in this case, the decimal for $1/p$ has period length $p-1$.

3.9   *Existence of primitive roots

The existence of a primitive root for each prime p is a subtle theorem, because we do not know any uniform way to specify a primitive root as a function of p. The *least* primitive root, for example, seems to vary with p in a highly irregular way. All known proofs of the theorem get around this difficulty by showing *only the existence* of a primitive root without attempting to find it.

The proofs use Lagrange's polynomial congruence theorem from Section 3.5, that the number of solutions of an nth degree congruence is $\leq n$. The theorem is used to show that, when $n < p - 1$, the congruences $x^n \equiv 1$ (mod p), have too few solutions to include all the $p - 1$ incongruent numbers 1, 2, 3, ..., $p - 1$. Thus at least one of these numbers satisfies only $x^{p-1} \equiv 1$ (mod p), and hence is a primitive root.

Together with this theorem, we use the proof of Fermat's little theorem from Section 3.4, which shows that each $a \not\equiv 0$ (mod p) satisfies a congruence $x^n \equiv 1$ (mod p), where n divides $p - 1$. This yields the following proposition on the number of solutions of $x^n \equiv 1$ (mod p).

Solutions of $x^n \equiv 1$ (mod p). *The congruence $x^n \equiv 1$ (mod p) has at most $\varphi(n)$ solutions that are not solutions of a congruence $x^m \equiv 1$ (mod p) of lower degree.*

Proof. If a satisfies $x^n \equiv 1$ (mod p) but no congruence $x^m \equiv 1$ (mod p) of lower degree, then a is of order n. Then $1, a, a^2, \ldots, a^{n-1}$ are distinct solutions of $x^n \equiv 1$ (mod p) and hence, by Lagrange's polynomial congruence theorem, they are the *only* solutions of $x^n \equiv 1$ (mod p).

Moreover, a power a^i such that $\gcd(i, n) > 1$ satisfies the lower-degree congruence $x^{n/\gcd(i,n)} \equiv 1$ (mod p). Thus the number of solutions of $x^n \equiv 1$ (mod p) that do not satisfy lower-degree congruences $x^m \equiv 1$ (mod p) is at most the number of i that are relatively prime to n, that is, $\varphi(n)$. $\square$

Finally, to prove the existence of primitive roots, we use this proposition to show that there are not enough elements of orders $n < p - 1$ to account for the $p - 1$ elements $1, 2, \ldots, p - 1$. To shorten notation we use $a|b$ for "a divides b".

Existence of primitive roots. *Less than $p - 1$ of the elements $1, 2, \ldots, p - 1$ have orders $n < p - 1$, hence one of them is a primitive root.*

Proof. By the previous proposition, the total number of elements with orders $n < p - 1$ is no more than

$$\sum_{\substack{n|p-1 \\ n \neq p-1}} \varphi(n).$$

We can prove that this number is less than $p - 1$ by proving that

$$\sum_{n|p-1} \varphi(n) = p - 1.$$

In fact, it is true for any natural number N that

$$\sum_{n|N} \varphi(n) = N.$$

To see why, consider the N fractions $\frac{1}{N}, \frac{2}{N}, \ldots, \frac{N}{N}$. Each of these has a reduced form $\frac{n'}{n}$, where $\gcd(n', n) = 1$, obtained by dividing the top and bottom by their gcd. For each divisor n of N there are $\varphi(n)$ reduced forms $\frac{n'}{n}$, and distinct fractions $\frac{i}{N}$ and $\frac{j}{N}$ have distinct reduced forms. Therefore

$$N = \sum_{n|N} \varphi(n),$$

as required.   □

Exercise

Here is another way to prove the existence of primitive roots, again assuming Lagrange's polynomial congruence theorem.

3.9.1 Suppose that the nonzero elements mod p have maximum order $n < p - 1$. Show that this implies $x^n \equiv 1 \pmod{p}$ for all the $p - 1$ nonzero values of x, mod p, contrary to Lagrange's polynomial congruence theorem.

3.10   Discussion

The congruence concept was introduced by Gauss (1801), who was the first to recognize its value in simplifying arguments involving division with remainders, such as Fermat's little theorem and Wilson's theorem. For example, instead of having to say "p divides a^{p-1} with remainder 1", one can write $a^{p-1} \equiv 1 \pmod{p}$, which looks and behaves like an equation.

Indeed, the concept of *congruence class*, introduced by Dedekind (1857), allows the congruence

$$a \equiv b \quad (\text{mod } n)$$

to be replaced by an *actual* equation

$$n\mathbb{Z} + a = n\mathbb{Z} + b,$$

between objects, $n\mathbb{Z} + a = \{nk + a : k \in \mathbb{Z}\}$ and $n\mathbb{Z} + b = \{nk + b : k \in \mathbb{Z}\}$, that obey the rules of arithmetic. This was an important step toward modern algebraic thinking, though ahead of its time, because few mathematicians accepted the use of sets as mathematical objects until the 20th century.

Fermat's little theorem grew out of the special case $2^{p-1} \equiv 1 \pmod{p}$, discovered by Fermat in an investigation of perfect numbers and primes of the form $2^p - 1$. He actually stated the theorem in the equivalent form $2^p \equiv 2 \pmod{p}$, and proved it using properties of the binomial coefficients. Fermat used neither the modern binomial theorem

$$(a+b)^p = a^p + \binom{p}{1}a^{p-1}b + \binom{p}{2}a^{p-2}b^2 + \cdots \binom{p}{p-1}ab^{p-1} + b^p,$$

nor the formula

$$\binom{p}{k} = \frac{p(p-1)\cdots(p-k+1)}{k!},$$

but a similar proof is easily obtained from them. One simply notes that

- For $k \neq 1, p$ the integer $\binom{p}{k}$ has the prime factor p in its numerator but not in its denominator. Hence p divides $\binom{p}{k}$.

- Therefore, by the binomial theorem,

$$2^p = (1+1)^p = 1^p + \binom{p}{1} + \binom{p}{2} + \cdots + \binom{p}{p-1} + 1^p$$

$$\equiv 2 \quad (\text{mod } p)$$

since p divides each of $\binom{p}{1}, \binom{p}{2}, \ldots, \binom{p}{p-1}$.

The equivalent form, $a^p \equiv a \pmod{p}$, of Fermat's little theorem may then be obtained by induction on a, since

$$3^p = (2+1)^p$$

$$\equiv 2^p + 1^p \pmod{p} \quad \text{since } \binom{p}{1}, \binom{p}{2}, \ldots, \binom{p}{p-1} \equiv 0 \pmod{p}$$

$$\equiv 2 + 1 \pmod{p} \quad \text{since } 2^p \equiv 2 \pmod{p}$$

$$\equiv 3 \pmod{p}, \quad \text{and so on.}$$

Around 1750 Euler gave a proof of Fermat's little theorem that fore-shadows the proof of Lagrange's theorem (20 years before Lagrange's own proof, which itself was not expressed in terms of groups; the group concept was introduced around 1830 by Galois).

Given $a \not\equiv 0 \pmod{p}$, let $\{1, a, a^2, \ldots, a^{n-1}\}$ be the set of distinct powers of a (which we recognize as a group A). Euler then shows that the distinct sets $\{b, ba, ba^2, \ldots, ba^{n-1}\}$ for the various $b \not\equiv 0 \pmod{p}$ (which we recognize as the cosets of A) form a partition of the set $\{1, 2, \ldots, p-1\}$. Hence the order n of a, which is the size of each coset, divides $p - 1$. Euler used a similar argument to prove his generalization of Fermat's little theorem. More on the early history of Fermat's little theorem and Euler's theorem may be found in Weil (1984).

Primes of the form $x^2 + ny^2$ are an important thread in the history of number theory, and we return to them several times in this book. The case $n = 1$ originates with Diophantus (if not earlier, in the study of Pythagorean triples) and his remark on products of sums of squares that we discussed in Section 3.7. By 1640 Fermat had completely mastered this case by reducing it to the question of which *primes* are of the form $x^2 + y^2$, showing that they are precisely the primes of the form $4n + 1$ (together with the obvious exceptional prime 2). We do not know how he proved it, except that he used descent, which was also the method of the first known proof, by Euler (1755). By 1654 Fermat had similarly dealt with primes of the form $x^2 + 2y^2$ and $x^2 + 3y^2$. As we saw in Section 3.7, it is easy to show that certain congruence classes are *not* of the required form. More powerful methods are required to show that other congruence classes *are* of the required form. We pick up this story again in Chapter 6.

The partial success of congruence arguments with the forms $x^2 + y^2$, $x^2 + 2y^2$, and $x^2 + 3y^2$ is not simply good luck. It can be explained by a sweeping general principle discovered by Hasse (1923) and called the *Hasse-Minkowski principle*. The principle implies that the impossibility of certain values for quadratic forms $ax^2 + bxy + cy^2$ can always be verified by congruence arguments.

4

The RSA cryptosystem

PREVIEW

The commonest application of number theory, and perhaps the most ubiquitous application of any kind of advanced mathematics, is the RSA cryptosystem. In this chapter we describe the system and how it works, based on a few key ideas from previous chapters.

The only theoretical ideas required are those of inverses mod n, the Euler φ function, and the related Euler theorem $a^{\varphi(n)} \equiv 1 \pmod{n}$. Allied with this are two fundamental algorithms: the algorithm for computing binary numerals, and the Euclidean algorithm (in the version that gives the inverse of a, mod b).

Thanks to the binary numeral algorithm, exponentiation mod n is feasible for large exponents. A "message" (viewed as an integer m) is encrypted as m^e mod n for certain publicly known e and n; and decrypted by raising the result to a power d, inverse to e mod $\varphi(n)$. This makes decryption easy only for someone who knows $\varphi(n)$.

4.1 Trapdoor functions

The science of cryptography seeks methods for encoding or *encrypting* messages, and corresponding methods for decoding or *decrypting*. Typically, encryption uses a certain *key* number (which may have many digits) and the same number is used for decryption. Without the key, it is not possible to read encrypted messages, so the security of the system depends on the difficulty of finding the key. Two well-known methods of encryption (at opposite ends of the security spectrum) are the following.

Example 1. The Caesar cipher.

This method of encryption (thought to have been used by Julius Caesar) simply adds the same integer *key number* (mod 26) to each letter in the message (viewed as a number between 1 and 26, assuming the Roman alphabet is used).

For example, if the key number is 3 then the message

Go to Zagreb tomorrow

is encrypted as

Jr wr Cdjuhe wrpruurz

and the latter is decrypted by subtracting 3 (mod 26) from each letter.

The Caesar cipher has low security because there are only 26 possible keys. It does not take an opponent very long to find the correct one— simply by trying the keys 1, 2, 3, … until one of them produces an intelligible message.

Example 2. The one-time pad.

In this method the key is a long, random sequence $x_1 x_2 x_3 \ldots$ of numbers x_i, each between 1 and 26. The digit x_i is added (mod 26) to the ith letter of the message to produce the encrypted message, and the receiver similarly subtracts x_i (mod 26) to recover the message. Once a segment $x_1 x_2 x_3 \ldots x_n$ of the key has been used for a message it is "torn off the pad", that is, the next portion $x_{n+1} x_{n+2} \ldots$ is used for the next message.

The one-time pad is completely secure (short of actually capturing a copy of the key) because all sequences $x_1 x_2 x_3 \ldots$ are equally likely, and hence so are all messages. There is no point even trying to guess the key. However, the key needs to be extremely long, since each segment of it is used only once, and this is inconvenient in practice.

The dream of cryptography has always been ease of implementation (as in the Caesar cipher) combined with security (as in the one-time pad), or at least a compromise between the two: it should be *feasible* to encrypt the message, but not feasible (without a reasonably short key) to decrypt it. Throughout history, this dream has failed time and time again, but it was revived in the 1970s in the mathematically more precise form of *trapdoor functions.*

A trapdoor function is an operation that is easy to do but hard to undo, like falling through a trapdoor or scrambling eggs. But unlike these real

life examples, *a trapdoor function is supposed to be easy to undo with
the help of a "key"*. Such functions seem to exist in mathematics, and
the theory of *polynomial time computability* has been developed to discuss
them. Here we illustrate these concepts with the example most important
for cryptography.

If we take two large prime numbers, say

$$p_1 = 4575163$$

and

$$p_2 = 4093567,$$

then we can easily find their product

$$p_1 p_2 = 18728736276421$$

(even using the school method of multiplication, which takes around n^2
steps for a pair of n-digit numbers).

Yet if we give someone the number 18728736276421 and ask them to
find the factors, it will probably take around a million steps. This is because
no known method for finding a divisor of a $2n$-digit number is substantially
quicker than trying to divide it by all 10^n numbers of $\leq n$ digits.

Thus the function $f(p_1, p_2) = p_1 p_2$ of numbers p_1, p_2 can be computed
in "quadratic time" but the inverse process of factorization seems to require
"exponential time". (These concepts can be made completely precise by
formalizing the concept of computation, but an informal understanding of
computing will suffice for our purposes.)

The seemingly hard-to-reverse property of multiplication is the basis
of the most commonly used cryptographic method today, the RSA system.
The system is named after the initials of the three mathematicians who first
published the system in 1978: Rivest, Shamir, and Adleman. It consists of

- an *encryption function* $E(m)$ of messages m that involves the product
  $(p_1 - 1)(p_2 - 1)$, where p_1 and p_2 are two large primes,

- a *decryption function* $D(m)$ that involves the two primes p_1 and p_2
  separately.

The encryption function is easily computed from the message and the
"key" $k = (p_1 - 1)(p_2 - 1)$ but the decryption function is not: it seems to
require factorization of the key to extract the primes p_1 and p_2. Because of

the apparent difficulty of factorization *the key k can be made public*, making $E(m)$ easy for everybody to compute, while $D(m)$ is easy to compute only for those who know p_1 and p_2.

Thus $E(m)$ is apparently a trapdoor function. We have to say "apparently", because no one has yet proved the underlying claim that factorization is hard. In view of the enormous number of communications that use RSA—military, commercial and private—this is an extremely important question. Regardless of what its answer turns out to be, the influence of RSA on number theory alone is enough to justify a short chapter on the subject.

4.2   Ingredients of RSA

A user of RSA owns a couple of large prime numbers, p_1 and p_2. If p_1 and p_2 are of, say, 100 digits, then the product $p_1 p_2$ can be computed in around 100^2 steps by the ordinary school method of multiplication. The product $p_1 p_2$ then has a unique factorization into two smaller factors, namely p_1 and p_2, but no known method of finding them is substantially better than dividing the 200-digit number $p_1 p_2$ by most of the approximately 10^{100} numbers less than its square root.

Thus the user can safely reveal the product $n = p_1 p_2$ without revealing its factors p_1 and p_2.

The theoretical ingredients of the RSA cryptosystem are inverses mod k and Euler's theorem, which we already have. The only other result we need is

$$\varphi(p_1 p_2) = (p_1 - 1)(p_2 - 1) \quad \text{for } p_1, p_2 \text{ prime.} \tag{*}$$

To prove (*) we ask how many natural numbers $a < p_1 p_2$ there are with $\gcd(a, p_1 p_2) = 1$. The only a for which this is *not* the case are the $p_2 - 1$ multiples of p_1 and the $p_1 - 1$ multiples of p_2. These $(p_1 + p_2) - 2$ numbers are distinct because $p_1 p_2$ is the smallest natural number that is a multiple of both p_1 and p_2. Hence

$$\varphi(p_1 p_2) = p_1 p_2 - 1 - (p_1 + p_2) + 2$$
$$= p_1 p_2 - p_1 - p_2 + 1$$
$$= (p_1 - 1)(p_2 - 1). \qquad \qquad \square$$

Knowing the primes p_1 and p_2, the user of RSA can easily compute $n = p_1 p_2$ and $\varphi(n) = (p_1 - 1)(p_2 - 1)$.

The user also chooses an *encryption exponent e*, which can be any number with

$$\gcd(e, \varphi(n)) = 1,$$

for example, a prime $< \varphi(n)$. The numbers e and n are made public, so anyone may use them to send encrypted messages to the user.

The value of $\varphi(n)$, known only to the user, enables computation of the *decryption exponent d*, which is the inverse of e, mod $\varphi(n)$. As we know, the inverse is easily computed from e and $\varphi(n)$ by the Euclidean algorithm.

The mathematical core of the RSA system is the following proposition, proved in Section 4.4. *If d is the inverse of e, mod $\varphi(k)$, then $(m^e)^d \equiv m$ (mod n).* Here m is the message, encryption raises m to the power e, mod n, and decryption recovers m by raising the encrypted message to the power d, mod n.

Encryption and decryption are feasible because *exponentiation mod n is easy to compute.* We explain why in the next section. The key to the success of RSA is the presumed difficulty of factorization, which makes $\varphi(n)$ and d hard to compute for anyone who does not know the two primes p_1 and p_2.

Exercises

To become familiar with the RSA system, take the (unrealistically small) primes $p_1 = 7$ and $p_2 = 11$.

4.2.1 Explain why $e = 5$ is not a valid encryption exponent.

4.2.2 Show that $e = 13$ is a valid encryption exponent and compute the corresponding decryption exponent d using the Euclidean algorithm.

4.2.3 Show that $e = 61$ is also a valid encryption exponent, but unsatisfactory because $m^{61} \equiv m$ (mod 77) for all $m \not\equiv 0$ (mod 77).

Such accidents, where raising to the power e does not change the message, are rare with the large primes p_1 and p_2 used in practice. Still, it shows that there are some subtleties in the proper choice of encryption exponent.

4.3 Exponentiation mod n

The obvious method to compute m^k is to form $m \times m \times \cdots \times m$ (k factors), which involves $k - 1$ multiplications. Since RSA uses exponents k with around 100 digits, the number of multiplications in this method of exponentiation will be around 10^{100}, a hopelessly large number. Thus the first step towards efficient exponentiation is to drastically reduce the number of multiplications; hopefully to a number around the size of $\log k$, which is proportional to the number of digits in k. We saw how to do this in Section 1.5, using the binary numeral for k.

Example. Construction of m^{91}.

We compute in turn

$$m = 1 \times m$$
$$m^2 = m^2$$
$$m^5 = (m^2)^2 \times m$$
$$m^{11} = (m^5)^2 \times m$$
$$m^{22} = (m^{11})^2$$
$$m^{45} = (m^{22})^2 \times m$$
$$m^{91} = (m^{45})^2 \times m$$

The total number of multiplications is the number of squarings (one less than the number of binary digits in k) plus the number of multiplications by m (no more than the number of binary digits in k). Hence the total number of multiplications to compute m^k is no more than twice the number of binary digits in k, and the number of binary digits is at most $\log_2 k + 1$.

It is still not a good idea to compute m^k for a 100-digit number k, even though it takes only about 200 multiplications, because the numbers being multiplied will become astronomical in length.

What makes RSA feasible is that we do not need m^k but *only its remainder on division by n*. Because of this we can compute with remainders throughout, using the arithmetic of congruences. In particular, we need never multiply numbers larger than n, and this is what makes exponentiation mod n feasible. Even by the school method of multiplication (which is not the most efficient known), multiplication of two n-digit numbers takes around n^2 steps, hence for n around 100 the work required for a couple of hundred multiplications is easily handled by a computer.

Exercises

4.3.1 Check that the above example allows m^{91} to be computed using 10 multiplications (not counting $1 \times m$).

4.3.2 Compute the binary numeral for 89, and hence show that m^{89} can be computed using 9 multiplications.

4.4   RSA encryption and decryption

If the user's primes are p_1 and p_2, a message is written (using some simple translation of letters into numerals) as a natural number m less than the publicly known product $n = p_1 p_2$. If the actual message is larger than this, it is broken into sufficiently small chunks that are encrypted one by one.

As foreshadowed in Section 4.2, the *encrypted message* sent to the user is the remainder of m^e when divided by n, which we abbreviate as

$$m^e \bmod n$$

This is a natural notation for remainders and it will not lead to confusion because

$$r = m^e \bmod n \quad \Rightarrow \quad r \equiv m^e \pmod{n}.$$

The numbers e and n are made public after having been computed by the user from the primes p_1 and p_2: $n = p_1 p_2$ and e is relatively prime to n. It is feasible to compute $m^e \bmod n$, even though e and n may have hundreds of digits, by the repeated squaring method explained in the previous section.

The user receives the encrypted message $m^e \bmod n$ and raises it to the power d, mod n, where d is the inverse of e, mod n. The result is the original message m, because d is an inverse of e, mod $\varphi(n)$, and therefore

$$ed = 1 + k\varphi(n) \quad \text{for some } k.$$

Hence

$$
\begin{aligned}
m^{ed} &= m^{1+k\varphi(n)} \\
&= m \cdot (m^{\varphi(n)})^k \\
&\equiv m(1)^k \pmod{n} \\
&\quad \text{since } m^{\varphi(n)} \equiv 1 \pmod{n}, \text{ by Euler's theorem,} \\
&\equiv m \pmod{n}.
\end{aligned}
$$

As with encryption, it is computationally feasible to raise a number to the power d, mod n, provided d is known. The decryption exponent d can be feasibly computed by the user, who knows the factors p_1, p_2 of n. These enable the computation of $\varphi(n) = (p_1 - 1)(p_2 - 1)$, and then the computation of d as the inverse of e, mod $\varphi(n)$, by the Euclidean algorithm.

This computation of the inverse is feasible because the Euclidean algorithm is similar in speed to exponentiation mod n on numbers of similar size, as remarked in Section 3.4, and $\varphi(n)$ is indeed just a little smaller than n when $n = p_1 p_2$.

Exercises

Continuing the toy example of RSA with $p_1 = 7$, $p_2 = 11$ and encryption exponent $e = 13$:

4.4.1 Show that the message m is encrypted as $((m^2 \cdot m)^2)^2 \cdot m$ mod 77.

4.4.2 When $m = 7$ verify that the encrypted message is 35.

It is not guaranteed, however, that every message is disguised by the encryption process. This is obviously not the case for $m = 1$ and it can also happen for other values:

4.4.3 When $m = 12$ verify that the encrypted message is also 12.

4.4.4 Using the decryption exponent from Exercise 4.2.2, verify that decryption of 12 recovers the message 12.

4.4.5 Explain the results of Exercises 4.4.2 and 4.4.3 by showing that $12^6 \equiv 1$ (mod 77).

4.5   Digital signatures

Another use of RSA is to transmit a digital *signature*—a proof that the user is who he or she claims to be. For this purpose the user can demonstrate possession of knowledge that no one else could have, such as the personal decryption exponent d that goes with the public numbers e and n.

This can be demonstrated, *without revealing d*, by taking some well known message m and sending m^d mod n. This is a scrambled message that only the possessor of d can create. But all the world knows e and n, hence they can unscramble m^d mod n by raising it to the power e, mod n:

$$(m^d)^e = m^{ed} \equiv m \quad (\text{mod } n), \quad \text{as above.}$$

Since only m^d mod n can unscramble to the recognizable message m in this fashion, the world can rest assured that the sender is indeed the possessor of the secret number d.

4.6   Other computational issues

The security of RSA depends, in the first instance, on having a large supply of 100-digit primes. If only a handful of such primes were available, an opponent could break the system by trying all pairs of them as p_1, p_2 until a product $p_1 p_2$ equal to n is found. Fortunately this is not a problem: there are many large primes and it is computationally easy to find them.

Thus an opponent's real problem is to compute the decryption exponent d from the publicly known e and n.

Since d is inverse to e, mod $\varphi(n)$, and $\varphi(n) = (p_1 - 1)(p_2 - 1)$, this would be feasible if the factors p_1 and p_2 of n were known. In fact it has been shown to be feasible *only* if the factors of n are known, hence decryption will remain difficult as long as factorization remains difficult.

However, it is *not known* whether factorization is truly difficult. No feasible method of factorization is known but it has not been proved that no such method exists. A proof that there is no feasible method would answer the so-called "$P \neq NP$ question", for which a prize of $1,000,000 has been offered.

Roughly speaking, problems of type P (for "polynomial time") can be solved by short computations, like the problem of multiplication. Problems of type NP (for "nondeterministic polynomial time") have solutions that are *verifiable* by short computations, but which may take a long time to find in the first place. As we have seen, factorization is like this. $P \neq NP$ says there are problems that are hard to solve but whose solutions are easy to verify. No such problem has yet been proved to exist though many good candidates are known (for example, the factorization problem).

4.7   Discussion

In the mid-70s, when mathematicians became aware of problems with solutions that were apparently hard to find but easy to verify, it was proposed to use such problems in *public key cryptosystems*—systems where it was easy to encrypt a message but hard to decrypt without extra, secret, information.

The idea of trapdoor functions, and their application to public key cryptosystems, was first published by Diffie and Hellman (1976). They also proposed exponentiation mod n as a computationally feasible process that might be hard to reverse. The implementation of this idea in RSA was first published by Rivest *et al.* (1978) and it has since become the most commonly used public key system. Just recently it was revealed that the same system was also discovered a few years earlier, by Clifford Cocks in the UK. Because it was part of his work for British Intelligence, it was kept secret (though why this was any use after 1978 is hard to understand). For more on the history of public key cryptosystems see Yan (2000).

The basic premise of RSA, that factorization is hard, was shaken by a remarkable discovery of Shor (1994). Shor found that factorization can be done in polynomial time on a *quantum computer*. The catch is that quantum computers do not yet exist and perhaps never will. Nevertheless Shor's result throws a strange new light on the concept of computation.

In all existing computers the difficulty in factorization (and in many other *NP* problems) is that the space of possible answers is exponentially large relative to the question. For an n-digit number K there are around $10^{n/2}$ numbers less than $\sqrt{K}$, and to factorize K we cannot do much better than try all of them as potential divisors. Since one has to try many things one after the other, factorization by all known methods takes exponential time.

According to quantum theory, however, in the world of the atom *many things actually happen at the same time in the same place*. The hypothetical quantum computer harnesses this possibility to do many computations simultaneously, and in this way it can factorize numbers in polynomial time. We say "hypothetical" advisedly, since it is not known whether a stable computer can actually be built from atom-sized components.

5

The Pell equation

PREVIEW

The so-called *Pell equation* $x^2 - ny^2 = 1$ (wrongly attributed to Pell by Euler) is one of the oldest equations in mathematics and it is fundamental to the study of quadratic Diophantine equations. The Greeks studied the special case $x^2 - 2y^2 = 1$ because they realized that its natural number solutions throw light on the nature of $\sqrt{2}$. There is a similar connection between the natural number solutions of $x^2 - ny^2 = 1$ and $\sqrt{n}$ when n is any nonsquare natural number.

The irrationality of $\sqrt{n}$ when n is nonsquare causes strange behavior in the solutions of $x^2 - ny^2 = 1$. Nevertheless, the irrationality of $\sqrt{n}$ reflects light back on the equation: it leads to simple algebraic structure, and a simple general formula for all integer solutions of $x^2 - ny^2 = 1$ in terms of the smallest natural number solution.

But there is *no* simple formula for the smallest natural number solution and it is not trivial even to prove that it exists. In this chapter we give two proofs: the first is a relatively direct proof due to Dirichlet, based on the approximation of $\sqrt{n}$ by rational numbers. The second (in the starred sections at the end of the chapter) is based on a more general theory of quadratic forms due to Conway.

We include Conway's theory because it is a natural extension of our study of the Euclidean algorithm (particularly the results in the starred sections of Chapter 2) and because it gives a very simple explanation of *periodicity* phenomena connected with the Pell equation and $\sqrt{n}$. It also gives a highly *visual* approach to the subject, which makes the complex behavior of the Pell equation surprisingly easy to grasp.

5.1   Side and diagonal numbers

The ancient Greeks met the equation $x^2 - 2y^2 = 1$ in their efforts to understand $\sqrt{2}$, the diagonal of the unit square, which they knew to be irrational.
They found a way to produce arbitrarily large solutions $(x_1, y_1), (x_2, y_2), \ldots$
of this equation, and hence fractions x_i/y_i that approximate $\sqrt{2}$ arbitrarily
closely. The fractions x_i/y_i tend to $\sqrt{2}$, because if $x_i^2 - 2y_i^2 = 1$ then

$$\frac{x_i^2}{y_i^2} = 2 + \frac{1}{y_i^2} \to 2 \quad \text{as } y_i \to \infty.$$

Thus if y_i is the side of a square, x_i approximates the diagonal.

The Greeks discovered the solutions (x_i, y_i) among the "side numbers"
s_i and "diagonal numbers" d_i defined by

$$d_1 = 3, \quad s_1 = 2,$$
$$d_{i+1} = d_i + 2s_i, \quad s_{i+1} = d_i + s_i.$$

It follows from these equations that

$$d_1^2 - 2s_1^2 = 1, \quad d_{i+1}^2 - 2s_{i+1}^2 = -(d_i^2 - 2s_i^2).$$

Hence the odd-numbered pairs (d_1, s_1), (d_3, s_3), (d_5, s_5), $\ldots$ satisfy the
equation $x^2 - 2y^2 = 1$ while the rest satisfy $x^2 - 2y^2 = -1$.

The first equation is an example of a *Pell equation*, the general form of
which is $x^2 - ny^2 = 1$ where n is a nonsquare integer. The second is closely
related to it; in fact we later look at *all* values of $x^2 - ny^2$ in order to see
whether they include the value 1.

Irrational square roots

In dealing with equations $x^2 - ny^2 = 1$, where n is a nonsquare integer, we
rely heavily on the irrationality of $\sqrt{n}$ proved in Section 2.5.

The upside of irrationality is that we can encode a pair of integers (a, b)
by a single real number $a + b\sqrt{n}$; we say that this number has *rational part*
a and *irrational part* b. Real and imaginary parts are meaningful because
if $\sqrt{n}$ is irrational, $a_1, b_1, a_2, b_2 \in \mathbb{Z}$, and

$$a_1 + b_1\sqrt{n} = a_2 + b_2\sqrt{n},$$

then $a_1 = a_2$ and $b_1 = b_2$.

Suppose, on the contrary, that $b_1 \neq b_2$. Then

$$a_1 - a_2 = (b_2 - b_1)\sqrt{n},$$

and, since $b_2 - b_1 \neq 0$, we get $\sqrt{n} = \frac{a_1 - a_2}{b_2 - b_1}$. This contradicts the irrationality of $\sqrt{n}$. Hence $b_1 = b_2$, and therefore $a_1 = a_2$. $\square$

Exercises

In the sections that follow we use numbers of the form $x_i + y_i\sqrt{n}$ to encode solution pairs of $x^2 - ny^2 = 1$. To give a taste of how this works, the following two exercises use numbers of the form $a + b\sqrt{2}$ to encode (diagonal,side) pairs.

5.1.1 Check that $(1 + \sqrt{2})^2 = 3 + 2\sqrt{2}$ and that

$$(x + y\sqrt{2})(1 + \sqrt{2}) = x + 2y + (x + y)\sqrt{2}.$$

5.1.2 Use induction to show from Exercise 5.1.1 that $(1 + \sqrt{2})^{n+1} = d_n + s_n\sqrt{2}$.

When n is an integer square, the equation $x^2 - ny^2 = 1$ is not so interesting, so we dispose of it right now.

5.1.3 By factorizing the left-hand side of $x^2 - y^2 = 1$, show that it has only two integer solutions.

5.1.4 Show similarly that $x^2 - ny^2 = 1$ has only two integer solutions when n is a square positive integer.

5.2   The equation $x^2 - 2y^2 = 1$

It is straightforward to find all *rational* solutions of $x^2 - ny^2 = 1$ by Diophantus' method (draw the line of slope t through the rational point $(1,0)$). Thus the method of solution is completely independent of n.

It is a different matter to find even one *integer* solution of $x^2 - ny^2 = 1$ other than the obvious ones $(\pm 1, 0)$. The least positive solution $\neq (\pm 1, 0)$ depends on n in a mysterious way. However, once this least nontrivial solution is found, all other integer solutions are generated by a simple formula. We illustrate the method for the case $n = 2$.

When $x^2 - 2y^2 = 1$ the smallest integer solution $\neq (\pm 1, 0)$ can be found by trial to be $(3, 2)$. Other solutions can then be found by the following *composition rule: if (x_1, y_1) and (x_2, y_2) are solutions of $x^2 - 2y^2 = 1$, then so is (x_3, y_3), where x_3 and y_3 are defined by*

$$(x_1 + y_1\sqrt{2})(x_2 + y_2\sqrt{2}) = x_3 + y_3\sqrt{2}.$$

To show that this rule gives a new solution we first calculate x_3 and y_3. Expanding the left-hand side, and collecting its rational and irrational parts, we find that

$$x_3 = x_1 x_2 + 2 y_1 y_2, \quad y_3 = x_1 y_2 + y_1 x_2.$$

It can then be checked by multiplication that

$$(x_1 x_2 + 2 y_1 y_2)^2 - 2(x_1 y_2 + y_1 x_2)^2 = (x_1^2 - 2 y_1^2)(x_2^2 - 2 y_2^2) = 1 \times 1 = 1.$$

Hence $x_3^2 - 2 y_3^2 = 1$, as required. $\square$

Examples. Composing the solution $(3, 2)$ with itself, we get a new solution (x_3, y_3), where

$$x_3 + y_3 \sqrt{2} = (3 + 2\sqrt{2})^2 = 9 + 8 + 12\sqrt{2} = 17 + 12\sqrt{2}.$$

Equating rational and irrational parts, $x_3 = 17$, $y_3 = 12$, which is indeed another solution. If we then compose $(17, 12)$ with $(3, 2)$ we get

$$(17 + 12\sqrt{2})(3 + 2\sqrt{2}) = 51 + 48 + (36 + 34)\sqrt{2} = 99 + 70\sqrt{2},$$

hence another solution is $(99, 70)$, and so on. By this process we can obtain infinitely many integer solutions, but it is not clear how close we are to finding all integer solutions. The situation becomes clearer when we observe that a *group structure* is present.

Exercises

Another way to arrive at the composition rule is to use the irrational factorization

$$x^2 - 2y^2 = (x - y\sqrt{2})(x + y\sqrt{2}). \tag{*}$$

We suppose that $1 = x_1^2 - 2 y_1^2$ and $1 = x_2^2 - 2 y_2^2$, so that

$$1 = 1 \times 1 = (x_1^2 - 2 y_1^2)(x_2^2 - 2 y_2^2). \tag{**}$$

5.2.1 Apply the factorization (*) to each factor on the right-hand side of (**), then combine the factors in a different way to show that

$$1 = [x_1 x_2 + 2 y_1 y_2 - (x_1 y_2 + y_1 x_2)\sqrt{2}]$$
$$\times [x_1 x_2 + 2 y_1 y_2 + (x_1 y_2 + y_1 x_2)\sqrt{2}].$$

5.2.2 Deduce from Exercise 5.2.1 that $x_3^2 - 2 y_3^2 = 1$, where

$$x_3 = x_1 x_2 + 2 y_1 y_2 \quad \text{and} \quad y_3 = x_1 y_2 + y_1 x_2.$$

In Section 5.4 we generalize this method to find a composition rule for solutions of $x^2 - ny^2 = 1$.

5.3   The group of solutions

Not only do solutions (x_1, y_1) and (x_2, y_2) of $x^2 - 2y^2 = 1$ have a "product" $(x_1 x_2 + 2y_1 y_2, x_1 y_2 + y_1 x_2)$, corresponding to the product of numbers

$$(x_1 + y_1 \sqrt{2})(x_2 + y_2 \sqrt{2}),$$

the numbers $x + y\sqrt{n}$ such that $x^2 - ny^2 = 1$ include $1 = 1 + 0\sqrt{n}$ and the multiplicative inverse $x - y\sqrt{2}$ of the number $x + y\sqrt{2}$:

$$(x + y\sqrt{2})(x - y\sqrt{2}) = x^2 - 2y^2 = 1$$

since $x^2 - 2y^2 = 1$ by the assumption that (x, y) is a solution.

Thus the solutions (x, y) form a *group*, with the same structure as the set of numbers $x + y\sqrt{2}$, where x, y are integers such that $x^2 - 2y^2 = 1$. To understand this group we first focus on the subgroup of *positive* numbers $x + y\sqrt{2}$ where $x^2 - 2y^2 = 1$.

Structure of positive solutions. *The group of positive $x + y\sqrt{2}$, where (x, y) is an integer solution of $x^2 - 2y^2 = 1$, is the infinite cyclic group of powers of $3 + 2\sqrt{2}$.*

To see why, apply the log function to all the positive numbers $x + y\sqrt{2}$ where x, y are integers such that $x^2 - 2y^2 = 1$. Since $\log(ab) = \log a + \log b$, the resulting numbers $\log(x + y\sqrt{2})$ then form a group under $+$.

This group has a least positive element, $\log(3 + 2\sqrt{2})$, because

- $3 + 2\sqrt{2}$ is the least $x + y\sqrt{2}$ corresponding to solutions (x, y) with $x, y > 0$,

- solutions $(x, -y)$ with $y > 0$ are inverses of solutions (x, y) with $x, y > 0$. Hence the corresponding $x - y\sqrt{2}$ are < 1, and their logs are < 0.

But any such group of numbers consists of the integer multiples of its least positive element m: if any element k lies between multiples of m,

$$mn < k < m(n + 1),$$

we also have $k - mn$ in the group, and the size of this element,

$$0 < k - mn < |m|,$$

contradicts the minimality of m. $\square$

Thus all solutions (x,y) of $x^2 - 2y^2 = 1$ for which $x + y\sqrt{2} > 0$ correspond to powers of $3 + 2\sqrt{2}$. Now for *any* solution (x,y) either $x + y\sqrt{2}$ or $-x - y\sqrt{2}$ is > 0. Hence the remaining solutions (x,y) are just the negatives of those obtained from the powers of $3 + 2\sqrt{2}$.

Exercises

Suppose we define integer pairs (u_k, v_k) by the equation

$$u_k + v_k\sqrt{2} = (3 + 2\sqrt{2})^k \quad \text{for all integers } k.$$

Then what we have just proved is that the pairs (u_k, v_k) are all the integer solutions (x,y) of $x^2 - 2y^2 = 1$ with x positive. It is now quite easy to express u_k and v_k as explicit functions of k, though (not surprisingly) these functions involve $\sqrt{2}$.

5.3.1 Given that $(3 + 2\sqrt{2})^k = u_k + v_k\sqrt{2}$, what is $(3 - 2\sqrt{2})^k$?

5.3.2 Deduce from Exercise 5.3.1 that

$$u_k = \frac{1}{2}\left[(3 + 2\sqrt{2})^k + (3 - 2\sqrt{2})^k\right], \quad v_k = \frac{1}{2\sqrt{2}}\left[(3 + 2\sqrt{2})^k - (3 - 2\sqrt{2})^k\right].$$

5.3.3 Deduce from Exercise 5.3.2 that $u_k = $ nearest integer to $(3 + 2\sqrt{2})^k/2$. And $v_k = ?$

5.4   The general Pell equation and $\mathbb{Z}[\sqrt{n}]$

If n is a nonsquare integer we define

$$\mathbb{Z}[\sqrt{n}] = \{x + y\sqrt{n} : x, y \in \mathbb{Z}\}.$$

Just as we used the numbers $x + y\sqrt{2}$ to study $x^2 - 2y^2 = 1$ we use the numbers $x + y\sqrt{n}$ to study $x^2 - ny^2 = 1$.

In fact, $x^2 - ny^2$ is what we call the *norm* of $x + y\sqrt{n}$ in $\mathbb{Z}[\sqrt{n}]$, the product of $x + y\sqrt{n}$ by its *conjugate* $x - y\sqrt{n}$:

$$\text{norm}(x + y\sqrt{n}) = (x - y\sqrt{n})(x + y\sqrt{n}) = x^2 - ny^2.$$

Thus finding solutions of the Pell equation is the same as finding elements of $\mathbb{Z}[\sqrt{n}]$ with norm 1.

The advantage of searching in $\mathbb{Z}[\sqrt{n}]$, rather than among pairs (x,y) of integers, is that we can use algebra on numbers in $\mathbb{Z}[\sqrt{n}]$.

Brahmagupta composition rule. *If (x_1, y_1) and (x_2, y_2) are both solutions of the Pell equation $x^2 - ny^2 = 1$, then so is*

$$(x_3, y_3) = (x_1 x_2 + n y_1 y_2, x_1 y_2 + y_1 x_2).$$

This generalizes the "composition" rule used for $n = 2$ in Section 5.2 and it may be proved as follows, using factorization in $\mathbb{Z}[\sqrt{n}]$.

Since (x_1, y_1) and (x_2, y_2) are solutions,

$$x_1^2 - ny_1^2 = 1 = x_2^2 - ny_2^2.$$

Therefore

$$
\begin{aligned}
1 &= (x_1^2 - ny_1^2)(x_2^2 - ny_2^2) \\
&= (x_1 - y_1\sqrt{n})(x_1 + y_1\sqrt{n}) \times (x_2 - y_2\sqrt{n})(x_2 + y_2\sqrt{n}) \\
&= (x_1 - y_1\sqrt{n})(x_2 - y_2\sqrt{n}) \times (x_1 + y_1\sqrt{n})(x_2 + y_2\sqrt{n}) \\
&= [x_1 x_2 + ny_1 y_2 - (x_1 y_2 + y_1 x_2)\sqrt{n}] \times [x_1 x_2 + ny_1 y_2 + (x_1 y_2 + y_1 x_2)\sqrt{n}] \\
&= (x_1 x_2 + ny_1 y_2)^2 - n(x_1 y_2 + y_1 x_2)^2 \\
&= x_3^2 - ny_3^2 \qquad\qquad\qquad\qquad\qquad\qquad\qquad\qquad\qquad \square
\end{aligned}
$$

This "composition" of solutions to form a new solution was discovered by the Indian mathematician Brahmagupta around 600 CE (but without using $\sqrt{n}$).

We also have an identity solution $(1, 0)$ and an inverse $(x, -y)$ of each solution (x, y), hence the solutions form a group, as we saw previously in the special case $n = 2$. As in that case, we can prove that all solutions come from powers of the smallest positive solution.

Example. Solutions of $x^2 - 3y^2 = 1$.

We find by trial that the smallest positive solution is $(2, 1)$. Composing $(2, 1)$ with itself we get the solutions

$$(2 \times 2 + 3 \times 1 \times 1, 2 \times 1 + 1 \times 2) = (7, 4),$$
$$(2 \times 7 + 3 \times 1 \times 4, 2 \times 4 + 1 \times 7) = (26, 15),$$

and so on. These solutions correspond to the powers of $2 + \sqrt{3}$.

The calculation used to prove the Brahmagupta composition rule actually shows a more general property, which holds not only with integer coefficients x, y but also with *rational* coefficients, that is, quotients of integers. We use the symbol $\mathbb{Q}$ ("quotients") for the rational numbers and make the natural generalization of $\mathbb{Z}[\sqrt{n}]$ to

$$\mathbb{Q}[\sqrt{n}] = \{x + y\sqrt{n} : x, y \in \mathbb{Q}\}.$$

This set of numbers is the set of quotients of elements of $\mathbb{Z}[n]$ and it is a number *field*, that is, closed under $+$, $-$, $\times$, and $\div$ (by nonzero members). The closure properties are easily checked by calculation (exercises).

We extend the definition of norm to $\mathbb{Q}[\sqrt{n}]$ by the same formula

$$\mathrm{norm}(x + y\sqrt{n}) = x^2 - ny^2.$$

This formula remains meaningful because each element of $\mathbb{Q}[\sqrt{n}]$ is uniquely expressible as $x + y\sqrt{n}$ with $x, y \in \mathbb{Q}$, by the argument of Section 5.1.

Multiplicative property of the norm. *For any α and β in $\mathbb{Q}[\sqrt{n}]$*

$$\mathrm{norm}(\alpha)\mathrm{norm}(\beta) = \mathrm{norm}(\alpha\beta).$$

Proof. Let $\alpha = x_1 + y_1\sqrt{n}$ and $\beta = x_2 + y_2\sqrt{n}$. Then

$$\mathrm{norm}(\alpha)\mathrm{norm}(\beta) = (x_1^2 - ny_1^2)(x_2^2 - ny_2^2)$$
$$= (x_1x_2 + ny_1y_2)^2 - n(x_1y_2 + y_1x_2)^2$$
$$\text{by the calculation above}$$
$$= \mathrm{norm}(\alpha\beta). \qquad \square$$

Exercises

5.4.1 Show that $+$, $-$, and $\times$ of numbers in $\mathbb{Q}[n]$ are themselves numbers in $\mathbb{Q}[n]$.

5.4.2 Show that $1/(x + y\sqrt{n})$ for $x, y \in \mathbb{Q}$ (not both zero) is of the form $x' + y'\sqrt{n}$ for $x', y' \in \mathbb{Q}$. Deduce that $\mathbb{Q}[n]$ is closed under $\div$ by nonzero members.

The multiplicative property of the norm can be restated as follows.

5.4.3 If (x_1, y_1) satisfies $x^2 - ny^2 = k_1$ and (x_2, y_2) satisfies $x^2 - ny^2 = k_2$, show that $(x_1x_2 + ny_1y_2, x_1y_2 + y_1x_2)$ satisfies $x^2 - ny^2 = k_1k_2$.

Brahmagupta used this fact to solve $x^2 - ny^2 = 1$ via easier equations $x^2 - ny^2 = k$. His method is most convenient when there is an obvious solution of $x^2 - ny^2 = -1$.

5.4.4 Find a nontrivial solution of $x^2 - 17y^2 = -1$ by inspection, and use it to find a nontrivial solution of $x^2 - 17y^2 = 1$.

5.4.5 Similarly find a nontrivial solution of $x^2 - 37y^2 = 1$.

5.5   The pigeonhole argument

The smallest nontrivial solution of $x^2 - ny^2 = 1$ is not always so easy to find as for $n = 2$ and $n = 3$. For example, the smallest nontrivial solution of $x^2 - 61y^2 = 1$ is

$$(x, y) = (1766319049, 226153980)!$$

This amazing example was discovered by Bhaskara II in 12th century India and rediscovered by Fermat.

The smallest nontrivial solution appears so unpredictably that its existence is not clear in general. However, Lagrange proved in 1768 that *if n is any nonsquare positive integer, the Pell equation $x^2 - ny^2 = 1$ has an integer solution $\neq (\pm 1, 0)$.*

An interesting new proof of this was given by Dirichlet around 1840. He used what is now called the "pigeonhole principle": if more than k pigeons go into k boxes then at least one box contains at least two pigeons (finite version); if infinitely many pigeons go into k boxes, then at least one box contains infinitely many pigeons (infinite version).

Dirichlet's argument can be subdivided into the following steps. First, a theorem on the approximation of irrational numbers:

Dirichlet's approximation theorem. *For any irrational $\sqrt{n}$ and integer $B > 0$ there are integers a, b with $0 < b < B$ and*

$$|a - b\sqrt{n}| < \frac{1}{B}.$$

Proof. For any integer $B > 0$ consider the $B - 1$ numbers $\sqrt{n}$, $2\sqrt{n}$ ..., $(B-1)\sqrt{n}$. For each multiplier k choose the integer A_k such that

$$0 < A_k - k\sqrt{n} < 1.$$

Since $\sqrt{n}$ is irrational, the $B - 1$ numbers $A_k - k\sqrt{n}$ are strictly between 0 and 1 and they are all different for the same reason (by the result of Section 5.1). Thus we have $B + 1$ different numbers

$$0, \quad A_1 - \sqrt{n}, \quad A_2 - 2\sqrt{n}, \quad \ldots, \quad A_{B-1} - (B-1)\sqrt{n}, \quad 1$$

in the interval from 0 to 1.

If we then divide this interval into B subintervals of length $1/B$, it follows by the finite pigeonhole principle that at least one subinterval contains two of the numbers. The difference between these two numbers, which is of the form $a - b\sqrt{n}$ for some integers a and b, is therefore irrational and such that

$$|a - b\sqrt{n}| < \frac{1}{B}.$$

Also, $b < B$ because b is the difference of two positive integers less than B. □

The next few steps are short and directed towards applications of the infinite pigeonhole principle.

1. Since Dirichlet's approximation theorem holds for all $B > 0$, we can make $1/B$ arbitrarily small, thus forcing the choice of new values of a and b. Thus *there are infinitely many integer pairs (a,b) with* $|a - b\sqrt{n}| < 1/B$. Since $0 < b < B$, we have

$$|a - b\sqrt{n}| < \frac{1}{b}.$$

2. It follows from step 1 that

$$|a + b\sqrt{n}| \leq |a - b\sqrt{n}| + |2b\sqrt{n}| \leq |3b\sqrt{n}|,$$

and therefore

$$|a^2 - nb^2| \leq \frac{1}{b} \cdot 3b\sqrt{n} = 3\sqrt{n}.$$

Hence *there are infinitely many $a - b\sqrt{n} \in \mathbb{Z}[\sqrt{n}]$ with norm of size* $\leq 3\sqrt{n}$.

3. By the infinite pigeonhole principle we obtain in turn

   • infinitely many $a - b\sqrt{n}$ with the same norm, N say,
   • infinitely many of these with a in the same congruence class, mod N,
   • infinitely many of these with b in the same congruence class, mod N.

4. From step 3 we get two positive numbers, $a_1 - b_1\sqrt{n}$ and $a_2 - b_2\sqrt{n}$, with

   • the same norm N,

- $a_1 \equiv a_2 \pmod{N}$,

- $b_1 \equiv b_2 \pmod{N}$.

The final step uses the quotient $a - b\sqrt{n}$ of the two numbers just found. Its norm $a^2 - nb^2$ is clearly 1 by the multiplicative property of norm. It is not so clear that a and b are integers, but this now follows from the congruence conditions in step 4.

Nontrivial solution of the Pell equation. *When n is a nonsquare positive integer, the equation $x^2 - ny^2 = 1$ has an integer solution $(a,b) \neq (\pm 1, 0)$.*

Proof. Consider the quotient $a - b\sqrt{n}$ of the two numbers $a_1 - b_1\sqrt{n}$ and $a_2 - b_2\sqrt{n}$ found in step 4. We have

$$a - b\sqrt{n} = \frac{a_1 - b_1\sqrt{n}}{a_2 - b_2\sqrt{n}} = \frac{(a_1 - b_1\sqrt{n})(a_2 + b_2\sqrt{n})}{a_2^2 - nb_2^2}$$

$$= \frac{a_1 a_2 - n b_1 b_2}{N} + \frac{a_1 b_2 - b_1 a_2}{N}\sqrt{n},$$

where $N = a_2^2 - nb_2^2$ is the common norm of $a_1 - b_1\sqrt{n}$ and $a_2 - b_2\sqrt{n}$. Since the latter numbers have equal norms, their quotient $a - b\sqrt{n}$ has norm 1 by the multiplicative property of norm (Section 5.4).

Since $a_1 - b_1\sqrt{n}$ and $a_2 - b_2\sqrt{n}$ are unequal and positive, their quotient $a - b\sqrt{n} \neq \pm 1$. It remains to show that a and b are integers. This amounts to showing that N divides $a_1 a_2 - n b_1 b_2$ and $a_1 b_2 - b_1 a_2$, or that

$$a_1 a_2 - n b_1 b_2 \equiv a_1 b_2 - b_1 a_2 \equiv 0 \pmod{N}.$$

The first congruence follows from the fact that $a_1^2 - nb_1^2 = N$, which implies

$$0 \equiv a_1^2 - nb_1^2 \equiv a_1 a_1 - n b_1 b_1 \equiv a_1 a_2 - n b_1 b_2 \pmod{N},$$

replacing a_1 and b_1 by their respective congruent values $a_1 \equiv a_2 \pmod{N}$ and $b_1 \equiv b_2 \pmod{N}$ found in step 4.

The second congruence follows from $a_1 \equiv a_2 \pmod{N}$ and $b_2 \equiv b_1 \pmod{N}$ by multiplying to obtain $a_1 b_2 \equiv b_1 a_2 \pmod{N}$, in other words, $a_1 b_2 - b_1 a_2 \equiv 0 \pmod{N}$. $\qquad\square$

5.6 *Quadratic forms

Dirichlet's pigeonhole argument is one of the neatest ways to prove the existence of nontrivial solutions of the Pell equation and it contains ideas that can be applied in other situations. Nevertheless, it is not obviously relevant to other quadratic Diophantine equations, so there is reason give a second proof: one that draws on a general theory of quadratic forms.

A *binary quadratic form* $Ax^2 + Bxy + Cy^2$, where $A, B, C \in \mathbb{Z}$, can be viewed as an integer-valued function of integer pairs, or *vectors* (x, y). Many classical questions in number theory are concerned with the values of quadratic forms. For example, the Pell equation asks whether 1 is a value of the form $x^2 - ny^2$, when n is a nonsquare natural number. To approach such questions we use two elementary properties of quadratic forms that can be confirmed by simple algebra.

Properties of quadratic forms. *If* $f(x, y) = Ax^2 + Bxy + Cy^2$ *and* $\mathbf{v} = (x, y)$ *then*

1. $f(k\mathbf{v}) = k^2 f(\mathbf{v})$,

2. $f(\mathbf{v}_1 + \mathbf{v}_2) + f(\mathbf{v}_1 - \mathbf{v}_2) = 2[f(\mathbf{v}_1) + f(\mathbf{v}_2)]$

Proof. 1. If $\mathbf{v} = (x, y)$ then $k\mathbf{v} = (kx, ky)$. Hence

$$f(k\mathbf{v}) = A(kx)^2 + B(kx)(ky) + C(ky)^2 = k^2(Ax^2 + Bxy + Cy^2) = k^2 f(\mathbf{v}).$$

2. If $\mathbf{v}_1 = (x_1, y_1)$ and $\mathbf{v}_2 = (x_2, y_2)$ then

$$f(\mathbf{v}_1) = Ax_1^2 + Bx_1y_1 + Cy_1^2 \quad \text{and} \quad f(\mathbf{v}_2) = Ax_2^2 + Bx_2y_2 + Cy_2^2.$$

Also

$$\begin{aligned} f(\mathbf{v}_1 + \mathbf{v}_2) &= A(x_1 + x_2)^2 + B(x_1 + x_2)(y_1 + y_2) + C(y_1 + y_2)^2 \\ &= Ax_1^2 + Ax_2^2 + Bx_1y_1 + Bx_2y_2 + Cy_1^2 + Cy_2^2 \\ &\quad + 2Ax_1x_2 + Bx_2y_1 + Bx_1y_2 + 2Cy_1y_2, \\ f(\mathbf{v}_1 - \mathbf{v}_2) &= A(x_1 - x_2)^2 + B(x_1 - x_2)(y_1 - y_2) + C(y_1 - y_2)^2 \\ &= Ax_1^2 + Ax_2^2 + Bx_1y_1 + Bx_2y_2 + Cy_1^2 + Cy_2^2 \\ &\quad - 2Ax_1x_2 - Bx_2y_1 - Bx_1y_2 - 2Cy_1y_2. \end{aligned}$$

Hence

$$f(\mathbf{v}_1+\mathbf{v}_2)+f(\mathbf{v}_1-\mathbf{v}_2) = 2Ax_1^2+2Bx_1y_1+2Cy_1^2+2Ax_2^2+2Bx_2y_2+2Cy_2^2$$
$$= 2[f(\mathbf{v}_1)+f(\mathbf{v}_2)] \qquad \square$$

A simple consequence of Property 1 is that $f(-\mathbf{v}) = f(\mathbf{v})$, so a quadratic form makes no distinction between a vector $\mathbf{v}$ and its negative. Property 1 also says that $f(k\mathbf{v})$ is a multiple of $f(\mathbf{v})$; in particular $f(\mathbf{v})$ is *prime* (or 1) only for vectors $\mathbf{v} = (x,y)$ that are not integer multiples of other integer vectors, that is, for (x,y) with relatively prime x and y. We call these *primitive vectors*.

In Section 2.8 we found a map of all the primitive vectors with positive x and y. We also found that the latter vectors are generated from $\mathbf{i} = (1,0)$ and $\mathbf{j} = (0,1)$ by the processes $(\mathbf{v}_1,\mathbf{v}_2) \mapsto (\mathbf{v}_1+\mathbf{v}_2,\mathbf{v}_2)$ and $(\mathbf{v}_1,\mathbf{v}_2) \mapsto (\mathbf{v}_1,\mathbf{v}_1+\mathbf{v}_2)$. In the next section we see that vectors with x and y of opposite sign are similarly generated from $(0,-1)$ and $(1,0)$. Then Property 2 shows that there is a simple relation between the values of f at successive stages in these processes. This leads to a "map" of the values of f.

Equivalent forms

Another view of a quadratic form f, related to the one described above, surveys all *equivalent* forms $f^*(x,y) = f(px+qy, rx+sy)$, obtained by replacing the row vector $(x\ y)$ by

$$(px+qy\ \ rx+sy) = (x\ y)\begin{pmatrix} p & r \\ q & s \end{pmatrix} = (x\ y)M,$$

where the matrix M and its inverse M^{-1} both have integer entries. When M satisfies these conditions, the pairs $(px+qy, rx+sy)$ run through the set $\mathbb{Z}^2$ of all integer pairs when (x,y) does. Indeed, if (x',y') is any integer pair, we have

$$(x'\ y') = (x\ y)M \Leftrightarrow (x\ y) = (x'\ y')M^{-1}.$$

Thus equivalent forms have the same set of values. Examples are x^2+y^2 and $x^2+2xy+2y^2$, the latter obtained from x^2+y^2 when (x,y) is replaced by $(x+y,y)$.

When M and M^{-1} both have integer entries, then $\det M$ and $\det M^{-1}$ are both integers. Since

$$MM^{-1} = \begin{pmatrix} 1 & 0 \\ 0 & 1 \end{pmatrix},$$

it follows by taking the determinant of both sides that

$$\det M \cdot \det M^{-1} = 1$$

(due to the multiplicative property: $\det(M_1 M_2) = \det M_1 \cdot \det M_2$). The only possible values for $\det M$ and $\det M^{-1}$ are therefore ± 1. Thus the condition for a matrix M to define an equivalence of quadratic forms is that M have integer entries and that $\det M = ps - qr = \pm 1$. Such a matrix is called *unimodular*.

Now an arbitrary quadratic form can be expressed as a matrix product,

$$Ax^2 + Bxy + Cy^2 = (x \ y) \begin{pmatrix} A & B/2 \\ B/2 & C \end{pmatrix} \begin{pmatrix} x \\ y \end{pmatrix}. \qquad (*)$$

So it follows from what we have just seen that any equivalent form is obtained by replacing

$$\begin{pmatrix} A & B/2 \\ B/2 & C \end{pmatrix} \quad \text{by} \quad M \begin{pmatrix} A & B/2 \\ B/2 & C \end{pmatrix} M^{-1},$$

where M is unimodular. This is so because the new matrix effects the replacement of $(x \ y)$ by $(x \ y)M$.

Formula (*) reveals an *invariant* of the form $Ax^2 + Bxy + Cy^2$ under equivalence, namely the determinant $AC - B^2/4$ of its matrix. Indeed, the determinant of any equivalent,

$$\det M \begin{pmatrix} A & B/2 \\ B/2 & C \end{pmatrix} M^{-1},$$

is equal (again by the multiplicative property of determinants) to

$$\det M \det \begin{pmatrix} A & B/2 \\ B/2 & C \end{pmatrix} \det M^{-1} = (\pm 1)^2 \det \begin{pmatrix} A & B/2 \\ B/2 & C \end{pmatrix}$$

$$= \det \begin{pmatrix} A & B/2 \\ B/2 & C \end{pmatrix},$$

since $\det M = \det M^{-1} = \pm 1$ by hypothesis. Thus *all equivalents of the form* $Ax^2 + Bxy + Cy^2$ *have the same determinant.*

Exercises

Although equivalent forms have the same determinant, the converse is not always true. It so happens that the form $x^2 + y^2$ is equivalent to all other forms with determinant 1, but $x^2 + 5y^2$ is *not* equivalent to all other forms with determinant 5.

5.6.1 Show that $13x^2 + 16xy + 5y^2$ has determinant 1, and that it is equivalent to $x^2 + y^2$ via the matrix $M = \begin{pmatrix} 2 & 3 \\ 1 & 2 \end{pmatrix}$.

5.6.2 Show that $2x^2 + 2xy + 3y^2$ has the same determinant as $x^2 + 5y^2$, but that it is not equivalent to $x^2 + 5y^2$, by showing that $x^2 + 5y^2$ does not take the value 7.

5.6.3 More generally, show that $x^2 + 5y^2$ takes no values $\equiv 3$ or $7 \pmod{20}$, by working out the possible values of $x^2 + 5y^2 \pmod{20}$.

5.7   *The map of primitive vectors

In Section 2.8 we described a partition of the plane (a "map") into regions labelled by $(1,0)$, $(0,1)$ and all the primitive vectors (a,b) of natural numbers. Figure 5.1 (right half) shows this map again, rotated through $90°$, together with a near mirror image of it (left half) in which the second coordinate of each pair has a negative sign.

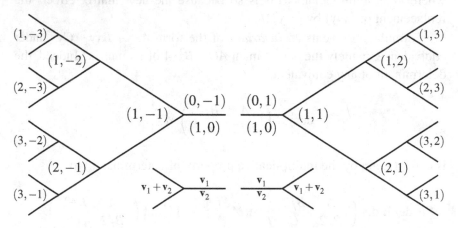

Figure 5.1: Two partial maps of primitive vectors

Also in the right half of the figure we have the schematic vector sum rule that generates all the labels from $(1,0)$ and $(0,1)$, and in the left half the mirror image rule that obviously applies there.

We put these two maps side by side because we want to join them together, but we seem prevented from doing so by the incompatible labels, $(0, 1)$ and $(0, -1)$, in the upper central region. The conflict can be resolved by giving each label a $\pm$ sign. This yields Figure 5.2, which we call the (complete) *map of primitive vectors*, for the obvious reason that it contains every primitive vector. The $\pm$ labelling fuses the two vector sum rules into the single *vector difference/sum rule* shown at the bottom of the Figure.

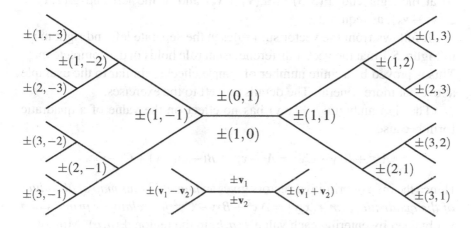

Figure 5.2: The complete map of primitive vectors

This rule needs some clarification because of the ambiguous signs. In a $\pm$ pair of vectors, say $\pm(1, 2)$, we are free to choose either $(1, 2)$ or $-(1, 2)$ as $\mathbf{v}_1$. Likewise for the pair, say $\pm(2, 3)$, labelling a region below an edge of region $\pm \mathbf{v}_1$: we can choose either $(2, 3)$ or $-(2, 3)$ to be $\mathbf{v}_2$. The vector difference/sum rule says that, *for some choice of* $\mathbf{v}_1$ *and* $\mathbf{v}_2$, the region between $\mathbf{v}_1$ and $\mathbf{v}_2$ at the left end of their common edge is labelled $\pm(\mathbf{v}_1 - \mathbf{v}_2)$ and the region at the right end is labelled $\pm(\mathbf{v}_1 + \mathbf{v}_2)$. In this example the regions are as in Figure 5.3.

$$\pm(1,2) \quad \pm(3,5)$$
$$= \quad \pm(1,1) \quad \pm(1,2) \quad \pm(3,5)$$
$$\pm(2,3) \qquad \pm(2,3)$$
$$\pm(1,1)$$

Figure 5.3: Regions above, below, and at the ends of an edge

Figure 5.3 shows how lines may be deformed to conform with the schematic diagram for the difference/sum rule—in particular the edge common to regions $\pm(1,2)$ and $\pm(2,3)$ is not really horizontal—within bounds that preserve the meanings of "above", "below", "right end", and "left end" for the edge common to the regions $\pm(1,2)$ and $\pm(2,3)$. Here the choice

$$\mathbf{v}_1 = (1,2), \; \mathbf{v}_2 = (2,3) \quad \text{gives} \quad \mathbf{v}_1 + \mathbf{v}_2 = (3,5), \; \mathbf{v}_1 - \mathbf{v}_2 = -(1,1),$$

so at the right end $\pm(3,5) = \pm(\mathbf{v}_1 + \mathbf{v}_2)$ and at the left end $\pm(1,1) = \pm(\mathbf{v}_1 - \mathbf{v}_2)$, as required.

It follows from the vector sum rules in the separate left and right maps in Figure 5.1 that the vector difference/sum rule holds in the complete map. This is proved by a finite number of simple checks, similar to the example above but more general. The details are left to the exercises.

The sign ambiguity $\pm(x,y)$ has no effect on the value of a quadratic form because

$$Ax^2 + Bxy + Cy^2 = A(-x)^2 + B(-x)(-y) + C(-y)^2.$$

Hence the map of primitive vectors gives an *unambiguous map of all values of the quadratic form* $f(x,y) = Ax^2 + Bxy + Cy^2$ *for relatively prime x and y*, obtained by entering each value $f(a,b)$ in the region $\pm(a,b)$. Moreover, it is possible to see some pattern in this map, thanks to the parallel between the vector difference/sum rule and Property 2 of quadratic forms proved in the previous section. We show this in the next section, assisted by the invariance of the determinant $AC - B^2/4$ under change of variables. The complete map also displays such changes, as we are about to see.

The tree of integral bases

In Section 5.6 we defined forms f, f^* to be *equivalent* if $f^*(x,y)$ results from $f(x,y)$ by replacing the vector (x,y) by a vector $(px+qy, rx+sy)$, which is equivalent to it in the sense that $(px+qy, rx+sy)$ runs through $\mathbb{Z}^2$ when (x,y) does. Since

$$(x,y) = x(1,0) + y(0,1) \quad \text{and} \quad (px+qy, rx+sy) = x(p,r) + y(q,s),$$

this amounts to replacing the vectors $(1,0)$ and $(0,1)$ by the new vectors (p,r) and (q,s). We call the pair of vectors $(1,0)$ and $(0,1)$ an *integral basis of* $\mathbb{Z}^2$ because any integer vector (x,y) is a linear combination of them with integer coefficients, namely $x(1,0) + y(0,1)$.

Equivalence says that the replacement $M : (x,y) \mapsto (px+qy, rx+sy)$ is invertible, so the inverse matrix M^{-1} has integer coefficients and the new vectors also form an integral basis. Thus the criterion for a pair of vectors (p,r) and (q,s) to form an integral basis is the criterion derived in Section 5.6 for M and M^{-1} to be integral, namely $ps - qr = \pm 1$.

Now in Section 2.7 we showed that, if (p,r) and (q,s) are labels on two regions with a common edge in the map of relatively prime pairs, then

$$ps - rq = \pm 1.$$

It is easily seen that this property extends to the complete map of Figure 5.2. Thus *each edge in the map of primitive vectors represents an integral basis of $\mathbb{Z}^2$*, namely the pair of labels on the regions that meet along the edge. The $\pm$ signs on the labels give four different bases, but they are essentially the same. Since the edges of the map form a tree, and each edge is associated in this way with an integral basis (up to sign), we call the edge complex of the map of primitive vectors the *tree of integral bases*.

As the name suggests, the tree represents *all* integral bases. We do not need this fact. However, it is easy to prove using the vector difference/sum rule to implement a kind of Euclidean algorithm (see exercises).

Exercises

To prove that the vector difference/sum rule holds in the complete map of primitive vectors we check that it holds in the middle and in "general position" on the right and left.

5.7.1 Verify that the difference/sum rule holds in the middle of the map (Figure 5.4) by choosing $\mathbf{v}_1 = (0,1)$ and $\mathbf{v}_2 = (1,0)$.

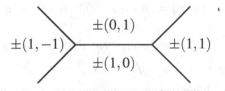

Figure 5.4: The middle of the complete map

5.7.2 Figure 5.5 shows one "general position" on the right side of the complete map. By choosing $\mathbf{v}_1 = \mathbf{u}_1$ and $\mathbf{v}_2 = \mathbf{u}_1 + \mathbf{u}_2$, verify that the difference/sum holds here.

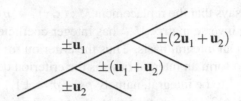

Figure 5.5: One "general position" on the right

5.7.3 Work out which other general positions occur on the right and on the left and verify that the difference/sum rule holds for each of them.

5.7.4 The "vector sum/difference rule" shown in Figure 5.6 is also valid. Why?

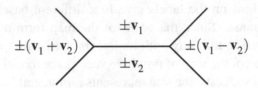

Figure 5.6: The sum/difference rule

To prove that the tree in the complete map represents all integral bases we use the difference/sum and sum/difference rules to trace a path from a given basis $\{(p,r),(q,s)\}$ back to $\{(1,0),(0,1)\}$. Exercise 5.7.5 is an example, and Exercises 5.7.6–5.7.8 show why such a path can always be found.

5.7.5 By repeatedly subtracting the "smaller" vector from the "larger", reduce the pair $\{(35,3),(23,2)\}$ to the pair $\{(1,0),(11,1)\}$. The latter pair is represented in the tree (why?), hence so is the former (why?).

5.7.6 Show that if
$$(p',r') = (p+q,r+s), \quad (q',s') = (q,s)$$
or
$$(p',r') = (p,r), \quad (q',s') = (p+q,r+s)$$
then $ps - qr = \pm 1 \Leftrightarrow p's' - q'r' = \pm 1$.

5.7.7 By repeatedly adding or subtracting one vector from the other, show that any pair $\{(p,r),(q,s)\}$ with $pr - qs = \pm 1$ reduces to a pair of the form $\{(p',0),(q',s')\}$. (*Hint*: $\gcd(r,s) = 1$. Why?) Deduce from Exercise 5.7.6 that $p' = \pm 1, q' = \pm 1$.

5.7.8 Deduce that $\{(p',0),(q',s')\}$ in Exercise 5.7.7 is represented by an edge in the tree, and hence so is $\{(p,r),(q,s)\}$.

5.8   *Periodicity in the map of $x^2 - ny^2$

In the previous section we briefly mentioned how a *map of any quadratic form f* may be superimposed on the map of primitive vectors by marking the region $\pm\mathbf{v}$ with the value $f(\mathbf{v}) = f(-\mathbf{v})$. We now investigate maps of quadratic forms in more depth and, to get an idea of what to expect, we first present the map of $x^2 - 3y^2$ in Figure 5.7. Only the right half is shown, because the left half is its mirror image. The values are marked as numbers in circles.

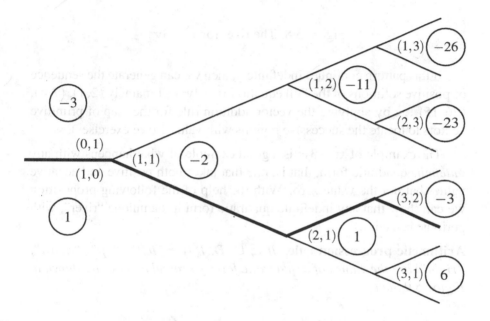

Figure 5.7: The map of $x^2 - 3y^2$

In this map there seems to be a single dividing line between positive and negative values of $x^2 - 3y^2$. Conway calls this line the *river*, and we have drawn it heavily in Figure 5.7. On either side of the river the values of $x^2 - 3y^2$ appear to increase in absolute value as one moves away from it (which is why one expects there to be only one river). And, rather unexpectedly, the values *along* the river seem to be *periodic*: in successive regions "above" the river the values are $-3, -2, -3, -2, \ldots$ and below each pair of successive regions with values $-3, -2$ there is a single region with value 1. Figure 5.8 confirms the pattern a bit further.

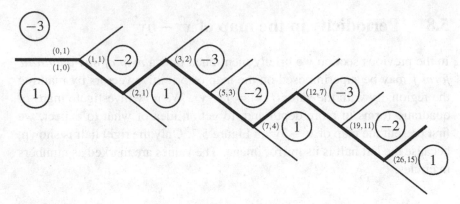

Figure 5.8: The river for $x^2 - 3y^2$

If this pattern continues indefinitely, then we can generate the sequence of positive solutions of the Pell equation $x^2 - 3y^2 = 1$, namely $(2,1)$, $(7,4)$, $(26,15),\dots$, by applying the vector addition rule for the map of primitive vectors to locate the successive regions with value 1 (see exercises).

The example of $x^2 - 3y^2$ is a good example of what happens with any *indefinite* quadratic form, that is, one that takes both positive and negative values but not the value zero. With the help of the following proposition we can show that any indefinite quadratic form has a unique "river", with periodic behavior.

Arithmetic progression rule. *If L, U, D, R (for "left", "up", "down", "right") are the values of a quadratic form f around an edge as shown in Figure 5.9 then*

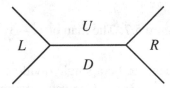

Figure 5.9: Values in regions around an edge

1. L, U + D, R is an arithmetic progression.

2. If (p,r) and (q,s) respectively are the regions above and below the edge, then the common difference in this progression is the coefficient of xy in the quadratic form $f(px + qy, rx + sy)$.

Proof. The difference/sum rule in the map of primitive vectors (Section 5.7) implies that

$$L = f(\mathbf{v}_1 - \mathbf{v}_2), \quad U = f(\mathbf{v}_1), \quad D = f(\mathbf{v}_2), \quad R = f(\mathbf{v}_1 + \mathbf{v}_2),$$

where $\mathbf{v}_1$ and $\mathbf{v}_2$ are the regions above and below the middle edge. It then follows from Property 2 of quadratic forms (Section 5.6) that

$$L + R = 2(U + D), \quad \text{or} \quad (U + D) - L = R - (U + D),$$

and this says that $L, U + D, R$ is an arithmetic progression.

Recall from Section 5.7 that if the basis $\mathbf{i} = (1, 0), \mathbf{j} = (0, 1)$ of $\mathbb{Z}^2$ is replaced by the basis $\mathbf{v}_1 = (p, r), \mathbf{v}_2 = (q, s)$, then the form $f(x, y)$ is replaced by the equivalent form $f^*(x, y) = f(px + qy, rx + sy) = Ax^2 + Bxy + Cy^2$ say. Also, the values of f at $\mathbf{v}_1, \mathbf{v}_2, \mathbf{v}_1 + \mathbf{v}_2$ and $\mathbf{v}_1 - \mathbf{v}_2$ are the same as the values f^* at $\mathbf{i}, \mathbf{j}, \mathbf{i} + \mathbf{j}$ and $\mathbf{i} - \mathbf{j}$, namely $A, C, A + B + C$ and $A - B + C$ respectively.

Thus the common difference, $(U + D) - L$, of the arithmetic progression is $A + C - (A - B + C) = B$, as claimed.   □

Part 1 of the arithmetic progression rule is enough to show:

Uniqueness of the river. *For any form $x^2 - ny^2$, where n is a nonsquare natural number, there is a unique edge path in the map of primitive vectors that separates regions of positive value from regions of negative value.*

Proof. Such a form is never zero, because $x^2 - ny^2 = 0$ implies $n = x^2/y^2$ is a square; and $x^2 - ny^2$ certainly takes both positive and negative values. Consider a place on its map where a region of value $L < 0$ meets two regions with values $U, D > 0$ as in Figure 5.9. (If the region with value L is actually on the right, it is still true that $L, U + D, R$ is an arithmetic progression.)

Then Part 1 implies that $R - (U + D) = (U + D) - L > U + D$, hence $R > \max(U, D)$. Thus moving one edge away from the border between positive and negative values leads to a region of greater positive value.

More generally, if $D > \max(U, L)$ then $R > D$ by a similar application of Part 1, so it follows that values of regions *continually increase* as we move further from the negative region. Similarly, values on the negative side continually decrease as we move further from the boundary path between positive and negative regions. Hence there is only one edge path separating the positive- from negative-valued regions.   □

We need Part 2 of the arithmetic progression rule to prove the more difficult periodicity property, which guarantees the existence of nontrivial solutions of the Pell equation.

Periodicity of the river. *When n is a nonsquare natural number, the pattern of values along the sides of the river for $x^2 - ny^2$ is periodic.*

Proof. It will suffice to prove that regions sharing edges with the river are bounded in absolute value. Indeed, if that is so, the values L, U and D in Figure 5.9 around some edge in the river will recur; hence so will the value R (being determined by L, U and D according to the arithmetic progression rule), whose region also shares an edge with the river, and so on.

As we saw in the proof of Part 2, the values U and D equal C and A, where $Ax^2 + Bxy + Cy^2$ is a quadratic form f^* equivalent to $f(x,y) = x^2 - ny^2$. But we know from Section 5.6 that the determinant $AC - B^2/4$ is the same for all equivalents f^* of f. Here C and A, being the values of regions on opposite sides of the river, have opposite signs. Hence

$$|AC - B^2/4| = |A||C| + B^2/4$$

Since $AC - B^2/4$ is constant, it follows that $|A|$ and $|C|$—the absolute values of D and U—are bounded as required.   □

Exercises

The "Pell quadratic forms" $x^2 - ny^2$ are by no means the only indefinite forms. Another interesting example is $x^2 + xy - y^2$, which is related to the *golden ratio* $\frac{1+\sqrt{5}}{2}$ and the Fibonacci sequence $1, 1, 2, 3, 5, 8, 13, \ldots$.

5.8.1 Show that $x^2 + xy - y^2 = \left(x + y\frac{1+\sqrt{5}}{2}\right)\left(x + y\frac{1-\sqrt{5}}{2}\right)$ and deduce from this that the form $x^2 + xy - y^2$ is indefinite.

5.8.2 Construct enough of the river for $x^2 + xy - y^2$ to show that its period looks like Figure 5.10.

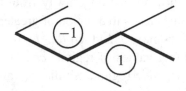

Figure 5.10: The period of $x^2 + xy - y^2$

5.8.3 Show that the positive labels (x_i, y_i) alternately below and above the river (in the regions marked alternately 1 and -1) satisfy

$$(x_1, y_1) = (1, 1), \quad (x_{i-1}, y_{i-1}) + (x_i, y_i) = (x_{i+1}, y_{i+1}).$$

5.8.4 Deduce from Exercise 5.8.3 that the natural number pairs satisfying the equation $x^2 + xy - y^2 = 1$ are (F_{2n+1}, F_{2n+2}) for $n = 0, 1, 2, 3, \ldots$, where $F_1 = F_2 = 1$ and $F_i + F_{i-1} = F_{i+1}$ (the Fibonacci sequence).

Periodicity in the shape of the river leads naturally to recurrence relations between the vectors labelling riverside regions. The Fibonacci relation arising from $x^2 + xy - y^2$ is the simplest example of such a recurrence relation. Another is the relation for $x^2 - 3y^2$, whose river was constructed above.

5.8.5 Use two successive periods in the river for $x^2 - 3y^2$ to show that the non-negative solutions (x_i, y_i) of $x^2 - 3y^2 = 1$ satisfy

$$(x_0, y_0) = (1, 0), \quad (x_{i+1}, y_{i+1}) = 4(x_i, y_i) - (x_{i-1}, y_{i-1}).$$

The river also shows why certain equations do *not* have solutions.

5.8.6 Explain why the equation $x^2 - 3y^2 = -1$ has no integer solution.

5.9 Discussion

The Pell equation $x^2 - ny^2 = 1$ is one of the oldest and most important quadratic Diophantine equations. Probably its only rival is the Pythagorean equation $x^2 + y^2 = z^2$. The Pell equation also dates back to the time of the Pythagoreans (around 500 BCE), who studied the special case $x^2 - 2y^2 = 1$ in connection with the $\sqrt{2}$, as mentioned in Section 5.1.

Another famous Pell equation is due to Archimedes. His "cattle problem" leads to the Pell equation $x^2 - 4729494y^2 = 1$, the least nontrivial solution of which has an x with 206545 digits! This solution was surely not known to Archimedes, though perhaps he knew that Pell equations could have remarkably large solutions. For an excellent discussion of the cattle problem, and the computational issues it raises, see Lenstra (2002).

The Pell equation was rediscovered in India, where mathematicians were also fascinated by short questions with long answers. Around 600 CE, Brahmagupta discovered the formula for composing solutions we used in Section 5.4. He used a generalization of it to find the minimal solution $(1151, 120)$ of $x^2 - 92y^2 = 1$ (saying that "a person solving this equation within a year is a mathematician"). In 1150 CE Bhaskara II extended Brahmagupta's idea to a method that solves all Pell equations, illustrating it with the well chosen example $x^2 - 61y^2 = 1$. He found its minimal solution, (1766319049,226153980), which is by far the largest minimal solution of any Pell equation $x^2 - ny^2 = 1$ with $n \leq 61$.

In Europe nothing was known of the Indian discoveries, but the Pell equation resurfaced in the 17th century when Fermat independently discovered how to solve it. He did not reveal his method, but he evidently knew what he was doing, because he too picked $x^2 - 61y^2 = 1$ as a challenge to other mathematicians. He also posed the even more formidable equation $x^2 - 109y^2 = 1$, the minimal solution of which is

$$(158070671986249, 15140424455100).$$

His English rivals Wallis and Brouncker rose to the challenge with a method that solves the Pell equation, not unlike the method of Bhaskara II (see Weil (1984), p. 94). In the 18th century these methods morphed into the simpler and more elegant *continued fraction algorithm*, which can be viewed as the Euclidean algorithm applied to the pair $(\sqrt{n}, 1)$.

All of these methods are based on the *observation of periodicity* in certain computations. It is likely that the ancient Greeks observed periodicity in the Euclidean algorithm, because simple geometric arguments show its periodicity on pairs such as $(\sqrt{2}, 1)$ and $(\sqrt{3}, 1)$ (see, for example, Stillwell (1998), p. 268, or Artmann (1999), p. 242). However, while many could *use* periodicity to solve instances of the Pell equation, the first to *prove* that periodicity always occurs was Lagrange (1768). He thereby showed that the continued fraction method always works. He underlined the importance of this result by showing that solving the Pell equation leads to the solution of *all* quadratic Diophantine equations in two variables.

Conway's visual approach, expounded in Sections 5.6–5.8, is certainly related to the old approaches to the Pell equation. But it is essentially simpler in that *it replaces a process (the Euclidean algorithm) by a picture (the map of primitive vectors)*. I have attempted to make this as clear as possible by deriving the map of primitive vectors and its properties directly from properties of the Euclidean algorithm, before imprinting the values of a quadratic form on it. (Conway assumes the simplest properties of the map, or sketches topological proofs, and proves others with the help of quadratic forms.) For further insights obtainable from Conway's approach, see the book Conway (1997) or his related video $ax^2 + hxy + by^2$ available from the American Mathematical Society.

6

The Gaussian integers

PREVIEW

The Gaussian integers $\mathbb{Z}[i]$ are the simplest generalization of the ordinary integers $\mathbb{Z}$ and they behave in much the same way. In particular, $\mathbb{Z}[i]$ enjoys *unique prime factorization*, and this allows us to reason about $\mathbb{Z}[i]$ the same way we do about $\mathbb{Z}$. We do this because $\mathbb{Z}[i]$ is the natural place to study certain properties of $\mathbb{Z}$. In particular, it is the best place to examine *sums of two squares*, because in $\mathbb{Z}[i]$ we can factorize a sum of two integer squares into linear factors: $x^2 + y^2 = (x - yi)(x + yi)$.

In the present chapter we use this idea to prove a famous theorem of Fermat: *if $p > 2$ is prime then $p = a^2 + b^2$, for some natural numbers a and b, if and only if $p = 4n + 1$ for some natural number n.* The Fermat two square theorem turns out to be related, not only to unique prime factorization in $\mathbb{Z}[i]$, but also to the actual "primes" of $\mathbb{Z}[i]$, the so-called *Gaussian primes*.

The Gaussian primes are easily shown to include the ordinary primes that are not sums of two squares, and the factors $a - bi$ and $a + bi$ of each ordinary prime of the form $a^2 + b^2$. Unique prime factorization in $\mathbb{Z}[i]$ establishes that these are the only Gaussian primes, up to multiples by ± 1 and $\pm i$.

An easy congruence argument shows that ordinary primes of the form $4n + 3$ are not sums of two squares. The two square theorem then shows that the primes that *are* sums of two squares are 2 and all the remaining odd primes, namely, those of the form $4n + 1$.

The proof of the two square theorem involves an important lemma proved with the help of Wilson's theorem: each prime $p = 4n + 1$

divides a number of the form $m^2 + 1$. Since $m^2 + 1$ factorizes in $\mathbb{Z}[i]$, it follows from unique prime factorization that p does also. The factorization of p turns out to be of the form $(a - bi)(a + bi)$, hence $p = (a - bi)(a + bi) = a^2 + b^2$, as claimed.

6.1 $\mathbb{Z}[i]$ and its norm

In the last chapter we saw that certain questions about $\mathbb{Z}$ are clarified by working with generalized integers, in particular, working in $\mathbb{Z}[\sqrt{n}]$ to solve $x^2 - ny^2 = 1$ in $\mathbb{Z}$. The role of $\mathbb{Z}[\sqrt{n}]$ in this case is to allow the factorization

$$x^2 - ny^2 = (x - y\sqrt{n})(x + y\sqrt{n}).$$

Similarly, when studying $x^2 + y^2$, it helps to use the *Gaussian integers*

$$\mathbb{Z}[i] = \{a + bi : a, b \in \mathbb{Z}\}$$

because $x^2 + y^2 = (x - yi)(x + yi)$.

Sums of two squares, $x^2 + y^2$, are the oldest known topic in number theory. We have already seen results about them found by the Babylonians, Euclid, and Diophantus. In fact, it could be said that some properties of $\mathbb{Z}[i]$ itself go back this far; at least, as far as Diophantus.

Diophantus apparently knew the two square identity (Section 1.8)

$$(a_1^2 + b_1^2)(a_2^2 + b_2^2) = (a_1 a_2 - b_1 b_2)^2 + (a_1 b_2 + b_1 a_2)^2$$

because he knew that the product of sums of two squares is itself the sum of two squares. Today we recognize this formula as equivalent to the *multiplicative property of absolute value*,

$$|z_1||z_2| = |z_1 z_2|,$$

where $z_1 = a_1 + b_1 i$ and $z_2 = a_2 + b_2 i$. And Diophantus' identity is exactly the formula

$$\mathrm{norm}(a_1 + b_1 i)\,\mathrm{norm}(a_2 + b_2 i) = \mathrm{norm}((a_1 + b_1 i)(a_2 + b_2 i)), \qquad (*)$$

where "norm" denotes the norm of $\mathbb{Z}[i]$,

$$\mathrm{norm}(a + bi) = |a + bi|^2 = a^2 + b^2.$$

Exercises

When discussing factorization there are always trivial factors, called *units*, that we prefer to ignore. For example, in $\mathbb{N}$ the only unit is 1, in $\mathbb{Z}$ the units are 1 and -1, and in $\mathbb{Z}[i]$ the units are the elements of norm 1.

6.1.1 Show that the units of $\mathbb{Z}[i]$ are $\pm 1, \pm i$.

Likewise, the units of $\mathbb{Z}[\sqrt{n}]$ are its elements of norm 1, that is, the numbers $a + b\sqrt{n}$ with $a^2 - nb^2 = 1$.

6.1.2 Describe the units of $\mathbb{Z}[\sqrt{2}]$.

6.1.3 Show that $\mathbb{Z}[\sqrt{n}]$ has infinitely many units for any nonsquare natural number n.

6.2 Divisibility and primes in $\mathbb{Z}[i]$ and $\mathbb{Z}$

The $\mathbb{Z}[i]$ norm

$$\operatorname{norm}(a + bi) = |a + bi|^2 = a^2 + b^2$$

is more useful in number theory than the absolute value because the norm is always an ordinary integer. The *multiplicative property of the norm* (*) implies that, if a Gaussian integer α divides a Gaussian integer γ, that is, if

$$\gamma = \alpha\beta \quad \text{for some } \beta \in \mathbb{Z}[i],$$

then

$$\operatorname{norm}(\gamma) = \operatorname{norm}(\alpha)\operatorname{norm}(\beta),$$

that is, $\operatorname{norm}(\alpha)$ *divides* $\operatorname{norm}(\gamma)$.

Because of this, questions about divisibility in $\mathbb{Z}[i]$ often reduce to questions about divisibility in $\mathbb{Z}$. In particular, it is natural to define a *Gaussian prime* to be a Gaussian integer that is not the product of Gaussian integers of smaller norm. Then we can answer various questions about Gaussian primes by looking at norms.

Examples.

1. $4 + i$ is Gaussian prime.
   Because $\operatorname{norm}(4 + i) = 16 + 1 = 17$, which is a prime in $\mathbb{Z}$. Hence $4 + i$ is not the product of Gaussian integers of smaller norm, because no such norms divide 17.

2. 2 is not a Gaussian prime.
   Because $2 = (1 - i)(1 + i)$ and both $1 - i$ and $1 + i$ have norm 2, which is smaller than $\text{norm}(2) = 4$.

3. $1 - i$, $1 + i$ are Gaussian prime factors of 2.
   Because $\text{norm}(1 - i) = \text{norm}(1 + i) = 2$ is a prime in $\mathbb{Z}$, hence $1 - i$ and $1 + i$ are not products of Gaussian integers of smaller norm.

Prime factorization in $\mathbb{Z}[i]$. *Any Gaussian integer factorizes into Gaussian primes.* The proof is similar to the proof in $\mathbb{Z}$.

Proof. Consider any Gaussian integer γ. If γ itself is a Gaussian prime, then we are done. If not, then $\gamma = \alpha\beta$ for some $\alpha, \beta \in \mathbb{Z}[i]$ with smaller norm. If α, β are not both Gaussian primes, we factorize into Gaussian integers of still smaller norm, and so on. This process must terminate since norms, being natural numbers, cannot decrease forever. Hence we eventually get a Gaussian prime factorization of γ.   □

As in $\mathbb{Z}$, it is not immediately clear that the prime factorization is unique. However, we see in Section 6.4 that unique prime factorization holds in $\mathbb{Z}[i]$ for much the same reasons as in $\mathbb{Z}$.

Exercises

An equivalent way to define Gaussian primes, in line with a common way of defining ordinary primes, is to say that ϖ is a Gaussian prime if ϖ is divisible only by units and units times ϖ. (It is conventional to use the Greek letter pi to denote primes in $\mathbb{Z}[i]$ and other generalizations of $\mathbb{Z}$, the way p is used to denote ordinary primes. However, to avoid confusion with $\pi = 3.14159\ldots$ I prefer to use ϖ, the variant form of pi.)

6.2.1 Explain why this definition is equivalent to the one above.

6.2.2 Prove that 3 is a Gaussian prime by considering the divisors of $\text{norm}(3)$.

Ordinary primes are not always Gaussian primes, as we have already seen in the case of 2. In fact, 2 is "almost a square" in $\mathbb{Z}[i]$.

6.2.3 Show that a unit times 2 is a square in $\mathbb{Z}[i]$.

6.2.4 Factorize 17 and 53 in $\mathbb{Z}[i]$.

6.3 Conjugates

The *conjugate* of $z = a + bi$ is $\bar{z} = a - bi$. The basic properties of conjugation (not only in $\mathbb{Z}[i]$ but for all complex numbers z) are

$$z\bar{z} = |z|^2,$$
$$\overline{z_1 + z_2} = \overline{z_1} + \overline{z_2},$$
$$\overline{z_1 - z_2} = \overline{z_1} - \overline{z_2},$$
$$\overline{z_1 \times z_2} = \overline{z_1} \times \overline{z_2}.$$

These can be checked by writing $z_1 = a_1 + b_1 i$, $z_2 = a_2 + b_2 i$ and working out both sides of each identity. We use these properties of conjugation to take the first step towards a classification of Gaussian primes.

Real Gaussian primes. *An ordinary prime $p \in \mathbb{N}$ is a Gaussian prime $\Leftrightarrow$ p is not the sum of two squares. (And obviously $p < 0$ is a Gaussian prime $\Leftrightarrow -p \in \mathbb{N}$ is a Gaussian prime.)*

Proof. ($\Leftarrow$) Suppose that we have an ordinary prime p that is not a Gaussian prime, so it factorizes in $\mathbb{Z}[i]$:

$$p = (a + bi)\gamma,$$

where $a + bi$ and γ are Gaussian integers with norm $<$ the norm p^2 of p (and hence also of norm > 1). Taking conjugates of both sides we get

$$p = (a - bi)\bar{\gamma},$$

since p is real and hence $p = \bar{p}$. Multiplying these two expressions for p gives

$$p^2 = (a - bi)(a + bi)\gamma\bar{\gamma}$$
$$= (a^2 + b^2)|\gamma|^2,$$

where both $a^2 + b^2, |\gamma|^2 > 1$. But the only such factorization of p^2 is pp, hence $p = a^2 + b^2$.

($\Rightarrow$) Conversely, *if an ordinary prime p equals $a^2 + b^2$ with $a, b \in \mathbb{Z}$ then p is not a Gaussian prime* because it has the Gaussian prime factorization

$$p = (a - bi)(a + bi)$$

into factors of norm $a^2 + b^2 = p < \text{norm}(p) = p^2$. $\qquad\square$

Notice also that *the factors $a - bi$ and $a + bi$ of p are Gaussian primes* because their norm is the prime number $a^2 + b^2 = p$. Moreover, all Gaussian primes $a + bi$, where $a, b \neq 0$, come in conjugate pairs like this. This is so because if one member of the pair factorizes into $\alpha\beta$ then its conjugate factorizes into $\overline{\alpha}\overline{\beta}$.

What is not yet clear is whether *all* Gaussian primes $a + bi$ with a, b nonzero are factors of ordinary primes $p = a^2 + b^2$. It is conceivable that $a + bi$ could be a Gaussian prime while $a^2 + b^2$ is a product of two or more ordinary primes. In Section 6.4 we rule this out with the help of unique prime factorization in $\mathbb{Z}[i]$.

At any rate, we can see that further clarification of the nature of Gaussian primes depends on finding another way to describe the ordinary primes that are sums of two squares. We saw in Section 3.7 (Example 1) that ordinary primes that are *not* sums of two squares are of the form $4n + 3$. The complement to this result—that any prime of the form $4n + 1$ *is* a sum of two squares—is a famous theorem discovered by Fermat. It is proved in Section 6.5.

Exercises

6.3.1 Verify the basic properties of conjugation mentioned above.

The proof of the classification of real Gaussian primes has the following interesting consequences.

6.3.2 Show that each ordinary prime has a distinct Gaussian prime associated with it.

6.3.3 Deduce that there are infinitely many Gaussian primes.

Since the real positive Gaussian primes are those of the form $4n + 3$, another way to prove that there are infinitely many Gaussian primes is to show that there are infinitely many ordinary primes of the form $4n + 3$. The proof is along lines similar to Euclid's proof in Section 1.1.

6.3.4 Show that the product of numbers of the form $4n + 1$ is of the same form. Deduce that any number of the form $4n + 3$ has a prime divisor of the form $4n + 3$.

6.3.5 If $p_1, p_2, \ldots, p_k$ are primes of the form $4n + 3$, show that $2p_1 p_2 \cdots p_k + 1$ is also of the form $4n + 3$.

6.3.6 Deduce from Exercises 6.3.4 and 6.3.5 that there are infinitely many primes of the form $4n + 3$.

6.4   Division in $\mathbb{Z}[i]$

Unique prime factorization in $\mathbb{Z}[i]$, as in $\mathbb{Z}$, relies on the Euclidean algorithm, which depends in turn on:

Division property of $\mathbb{Z}[i]$. *If $\alpha, \beta \neq 0$ are in $\mathbb{Z}[i]$ then there is a* quotient μ *and a* remainder ρ *such that*

$$\alpha = \mu\beta + \rho \quad \text{with} \quad |\rho| < |\beta|.$$

Proof. This property becomes obvious once one sees that the Gaussian integer multiples $\mu\beta$ of any Gaussian integer $\beta \neq 0$ form a square grid in the complex plane.

This is because multiplication of β by i rotates the vector from 0 to β through $90°$, hence 0, β, and $i\beta$ are three corners of a square. All other multiples of β are sums (or differences) of β and $i\beta$, hence they lie at the corners of a square grid. (Figure 6.1.)

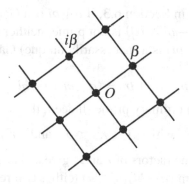

Figure 6.1: Multiples of a Gaussian integer

Any Gaussian integer α lies in one of these squares, and there is a nearest corner $\mu\beta$ (not necessarily unique, but no matter). Then

$$\alpha = \mu\beta + \rho, \quad \text{where} \quad |\rho| = \text{distance to nearest corner,}$$

so $|\rho|$ is less than the side of a square, namely $|\beta|$. $\square$

Thanks to the division property we have

1. A Euclidean algorithm for $\mathbb{Z}[i]$

2. $\gcd(\alpha, \beta) = \mu\alpha + \nu\beta$ for some $\mu, \nu \in \mathbb{Z}[i]$.

3. The *prime divisor property*: if a prime ϖ divides $\alpha\beta$ then ϖ divides α or ϖ divides β.

4. Unique prime factorization up to order and factors of norm 1, namely ± 1 and $\pm i$. Elements of norm 1 are called *units* and unique prime factorization usually comes with the qualification "up to unit factors". This is true even in $\mathbb{Z}$, where the units are ± 1 and hence primes may vary up to sign.

As a first application of unique prime factorization in $\mathbb{Z}[i]$ we complete the description of Gaussian primes begun in Section 6.3. There we found that the real Gaussian primes are ordinary primes that are not sums of two squares, and their negatives. It is also clear that the *pure imaginary* Gaussian primes are of the form $\pm ip$, where p is a real Gaussian prime. Thus it remains to describe the Gaussian primes $a + bi$ with a, b nonzero.

Imaginary Gaussian primes. *The Gaussian primes $a+bi$ with a, b nonzero are factors of ordinary primes p of the form $a^2 + b^2$.*

Proof. First, as noted in Section 6.3, if $a + bi$ is a Gaussian prime then so is $a - bi$ (because if $a - bi = \alpha\beta$ is not prime, neither is $a + bi = \overline{\alpha}\overline{\beta}$).

Next, $(a - bi)(a + bi)$ is a (necessarily unique) Gaussian prime factorization of

$$p = a^2 + b^2 = (a - bi)(a + bi).$$

But p must then be an ordinary prime. Indeed, if

$$p = rs \quad \text{with} \quad 1 < r, s < p \quad \text{and} \quad r, s \in \mathbb{Z},$$

then the Gaussian prime factors of r and s give a Gaussian prime factorization of p different from $(a - bi)(a + bi)$ (either two real factors r and s, or $\geq$ four complex factors). $\square$

Exercises

Using unique prime factorization we can prove results on squares and cubes in $\mathbb{Z}[i]$, similar to those on squares and cubes in $\mathbb{N}$ proved in Section 2.5. The only difference is that we have to take account of units, as indeed we already do in $\mathbb{Z}$.

6.4.1 Is it true in $\mathbb{Z}$ that relatively prime factors of a square are themselves squares? If not, how should the statement be modified to make it correct?

6.4.2 Show that relatively prime factors of a cube in $\mathbb{Z}$ are themselves cubes.

6.4.3 Formulate a theorem about relatively prime factors of a square in $\mathbb{Z}[i]$.

6.4.4 Show that relatively prime factors of a cube in $\mathbb{Z}[i]$ are themselves cubes.

6.5 Fermat's two square theorem

In Section 3.7 we used congruence mod 4 to show that primes of the form $4n + 3$ are not sums of two squares. Fermat's two square theorem says that the remaining odd primes—those of the form $4n + 1$—are all sums of two squares.

We apply the theory of $\mathbb{Z}[i]$ to a prime $p = 4n + 1$ with the help of an $m \in \mathbb{Z}$ such that p divides $m^2 + 1$. Such an m always exists by a result of Lagrange (1773) that follows from *Wilson's theorem* in Section 3.5: for any prime p

$$1 \times 2 \times 3 \times \cdots \times (p - 1) \equiv -1 \pmod{p}.$$

Lagrange's lemma. *A prime $p = 4n + 1$ divides $m^2 + 1$ for some $m \in \mathbb{Z}$.*

Proof. If we apply Wilson's theorem to the prime $p = 4n + 1$ we get

$$
\begin{aligned}
-1 &\equiv 1 \times 2 \times 3 \times \cdots \times 4n \pmod{p} \\
&\equiv (1 \times 2 \times \cdots \times 2n) \times \\
&\quad ((2n + 1) \times \cdots \times (4n - 1) \times (4n)) \pmod{p} \\
&\equiv (1 \times 2 \times \cdots \times 2n) \times \\
&\quad ((-2n) \times \cdots \times (-2)(-1)) \pmod{p} \qquad \text{since } p - k \equiv -k \pmod{p} \\
&\equiv (1 \times 2 \times \cdots \times 2n)^2 (-1)^{2n} \pmod{p} \\
&\equiv (1 \times 2 \times \cdots \times 2n)^2 \pmod{p}
\end{aligned}
$$

Taking $m = (2n)!$ we get $m^2 \equiv -1 \pmod{p}$. That is, p divides $m^2 + 1$. $\square$

Fermat's two square theorem. *If $p = 4n + 1$ is prime, then $p = a^2 + b^2$ for some $a, b \in \mathbb{Z}$.*

Proof. Given p, let $m \in \mathbb{Z}$ be such that p divides $m^2 + 1$, as in the lemma. In $\mathbb{Z}[i]$, $m^2 + 1$ has the factorization

$$m^2 + 1 = (m - i)(m + i).$$

And, even though p divides $m^2 + 1$, p does *not* divide $m - i$ or $m + i$ because $\frac{m}{p} - \frac{i}{p}$ and $\frac{m}{p} + \frac{i}{p}$ are not Gaussian integers.

By the Gaussian prime divisor property of Section 6.4, it follows that p is *not a Gaussian prime*. But then $p = a^2 + b^2$, as proved in Section 6.3. $\square$

It also follows that

$$p = (a - bi)(a + bi)$$

is a factorization into Gaussian primes, and we now know that any such factorization is unique. So in fact we have a stronger form of Fermat's two square theorem: *each prime $p = 4n + 1$ is a sum $a^2 + b^2$ of two squares for a unique pair of natural numbers a, b.*

Exercises

Here is another way in which $\mathbb{Z}[i]$ throws light on sums of two squares. The following exercises develop a proof of a theorem of Euler (1747): *if $\gcd(a,b) = 1$ then any divisor of $a^2 + b^2$ is of the form $c^2 + d^2$ where $\gcd(c,d) = 1$.* The main steps depend on unique prime factorization in $\mathbb{Z}[i]$.

6.5.1 Give an example that shows why the condition $\gcd(a,b) = 1$ is necessary.

6.5.2 Show that each integer divisor $e > 1$ of $a^2 + b^2$ is a product of Gaussian prime divisors $q + ir$ of $a^2 + b^2$, unique up to unit factors.

6.5.3 Show that each of the Gaussian primes $q + ir$ divides either $a - ib$ or $a + ib$. Deduce that none of them is an ordinary prime p.

6.5.4 Show that, along with each Gaussian prime factor $q + ir$ of e, its conjugate $q - ir$ is also a factor.

6.5.5 Deduce from Exercise 6.5.4 that e is of the form $c^2 + d^2$ where $c + di$ divides $a + bi$.

6.5.6 Deduce from Exercise 6.5.5 that $\gcd(c,d) = 1$.

6.6 Pythagorean triples

Now is a good time to revisit the primitive Pythagorean triples, whose relationship with $\mathbb{Z}[i]$ was suggested in Section 1.8. Since odd squares are congruent to 1 (mod 4) and even squares are congruent to 0 (mod 4), a sum of two odd squares is not a square. Hence in a primitive triple (x,y,z) one of x, y is even and z is odd. The argument in Section 1.8 was that if

$$x^2 + y^2 = z^2$$

then

$$(x - yi)(x + yi) = z^2,$$

so $x - yi$ and $x + yi$ are Gaussian prime factors of an odd square, z^2. Then we wanted to say that:

1. If x and y are relatively prime (in $\mathbb{Z}$) then so are $x - yi$ and $x + yi$ (in $\mathbb{Z}[i]$).

2. In $\mathbb{Z}[i]$, relatively prime factors of a square are squares.

The first statement is correct. If $\gcd(x,y) = 1$ in $\mathbb{Z}$ then $\gcd(x,y) = 1$ in $\mathbb{Z}[i]$. This is so since a common Gaussian prime divisor is accompanied by its conjugate, and their product is a common divisor >1 in $\mathbb{Z}$. A common divisor of $x - yi$ and $x + yi$ also divides their sum $2x$ and difference $2iy$. Therefore, since $\gcd(x,y) = 1$, any common prime divisors of $x - iy$ and $x + iy$ are primes $\pm 1 \pm i$ dividing 2. No such divisors are present, since they imply that $(x - iy)(x + iy) = z^2$ is even.

The second statement is not quite correct, but the following amendment of it is: *relatively prime factors of a square are squares, up to unit factors.* This follows from unique prime factorization in $\mathbb{Z}[i]$.

Since $x - yi$ and $x + yi$ have no common Gaussian prime factor, while each prime factor of z^2 occurs to an even power, each prime factor of $x - yi$, and each prime factor of $x + yi$, must also occur to an even power. A product of primes, each occurring to an even power, is obviously a square (compare with the same argument for natural numbers in Section 2.5). Hence each of $x - yi$ and $x + yi$ is a unit times a square, since their only possible nonprime factors are units. $\qquad\square$

The amended second statement is good enough to give us the conclusion we expect. We have shown that $x - yi$ is a unit times a square, hence it is one of

$$(s - ti)^2, \quad -(s - ti)^2, \quad i(s - ti)^2, \quad -i(s - ti)^2, \quad \text{for some } s, t \in \mathbb{Z}.$$

That is, it is one of

$$(s^2 - t^2) - 2sti, \quad t^2 - s^2 + 2sti, \quad 2st + (s^2 - t^2)i, \quad -2st + (t^2 - s^2)i.$$

In each case, equating real and imaginary parts gives one of x and y in the form $u^2 - v^2$ and the other in the form $2uv$ for some natural numbers u and v. Thus the result is essentially the same as that obtained by the loose argument in Section 1.8, but better, because it does not force the even member of the pair, $2uv$, to be first.

Moreover, we necessarily have $\gcd(u,v) = 1$ because any common prime divisor of u and v is a common divisor of $u^2 - v^2$ and $2uv$, hence of x and y. Thus the correct outcome of the speculation in Section 1.8 is:

Primitive Pythagorean triples. *If $x^2 + y^2 = z^2$ for some relatively prime natural numbers x and y, then one of x and y is of the form $u^2 - v^2$ and the other of the form $2uv$, for relatively prime natural numbers u and v.* □

We also find in each case that $z = u^2 + v^2$, because

$$(u^2 - v^2)^2 + (2uv)^2 = u^4 + 2u^2v^2 + v^4 = (u^2 + v^2)^2.$$

Thus z is a sum of two squares. Since u and v are any relatively prime numbers, and a prime $u^2 + v^2$ necessarily has $\gcd(u, v) = 1$, z can be any *prime* sum of two squares. Thus we get a geometric characterization of the primes that are sums of two squares.

Prime hypotenuses. *The primes that are sums of two squares are those that occur as hypotenuses of right-angled triangles with integer sides.* □

Exercises

The last result, together with Fermat's two square theorem, shows that the primes of the form $4n + 1$ are precisely those occurring as hypotenuses of integer right-angled triangles.

6.6.1 Find integer right-angled triangles with hypotenuses 5, 13, 17 (you should know these), and 29, 37, and 41.

6.6.2 Given a prime $p = 4n + 1$, is the integer right-angled triangle with hypotenuse p unique?

The argument above shows that, if (x, y, z) is a primitive Pythagorean triple, then $x + yi$ is a unit times a square in $\mathbb{Z}[i]$. But once we know that $x = u^2 - v^2$, $y = 2uv$ we can say more.

6.6.3 If (x, y, z) is a primitive Pythagorean triple with x odd, show that $x + yi$ is a square in $\mathbb{Z}[i]$.

6.6.4 Verify directly that $3 + 4i$ is a square in $\mathbb{Z}[i]$.

It should be clear from your answer to Question 6.6.3 that finding the parameters u and v for a given primitive Pythagorean triple (x, y, z), with x odd, is equivalent to finding the square root(s) of a complex number.

6.6.5 Find the square root of $5 + 12i$.

6.6.6 If you have some software for computing square roots of complex numbers, verify that each entry (x, y, z) in Plimpton 322 (Section 1.6), except the triple $(60, 45, 75)$, yields a $y + xi$ that is a square in $\mathbb{Z}[i]$. (Note: this includes the last triple $(90, 56, 106)$, which is clearly not primitive.)

6.6.7 Explain how to compute the square root of a complex number by hand, using quadratic equations.

6.7   *Primes of the form $4n+1$

Lagrange's lemma, proved in the Section 6.5, is actually half of an important result concerning the so-called "quadratic character of -1" that we study further in Chapter 9. Here we use it to prove that there are infinitely many primes of the form $4n+1$, complementing the corresponding easy result about primes of the form $4n+3$ proved in Exercises 6.3.4–6.3.6.

Quadratic character of -1. *The congruence* $x^2 \equiv -1$ *(mod p), where p is an odd prime, has a solution precisely when* $p = 4n+1$.

Proof. When $p = 4n+1$, Lagrange's lemma gives an x with $x^2 \equiv -1$ (mod p). To show that $x^2 \equiv -1$ (mod p) has no solution when $p = 4n+3$ we suppose, on the contrary, that it does.

If

$$x^2 \equiv -1 \quad (\bmod\ p = 4n+3)$$

then raising both sides to the power $2n+1$ gives

$$(x^2)^{2n+1} \equiv (-1)^{2n+1} \equiv -1 \quad (\bmod\ p = 4n+3).$$

Since $2(2n+1) = 4n+2 = p-1$, this says that

$$x^{p-1} \equiv -1 \quad (\bmod\ p),$$

contrary to Fermat's little theorem. Hence $x^2 \equiv -1$ (mod p) has no solution when $p = 4n+3$.   □

Thus solutions of $x^2 \equiv -1$ (mod p) occur precisely when the odd prime p is of the form $4n+1$. To put it another way: *the odd primes p that divide values of* $x^2 +1$, *for* $x \in \mathbb{Z}$, *are precisely the primes* $p = 4n+1$.

Infinitude of primes $4n+1$. *There are infinitely many primes of the form* $p = 4n+1$.

Proof. From what we have just proved, it suffices to show that infinitely many primes divide values of $x^2 +1$ for $x \in \mathbb{Z}$. Suppose on the contrary that only finitely many primes $p_1, p_2, \ldots, p_k$ divide values of $x^2 +1$.

Now consider the polynomial

$$(p_1 p_2 \cdots p_k y)^2 + 1 = g(y).$$

Clearly, any prime p that divides a value of $g(y)$, for $y \in \mathbb{Z}$, also divides a value of $x^2 +1$ (namely, for $x = p_1 p_2 \cdots p_k y$). But none of $p_1, p_2, \ldots, p_k$ divides $g(y)$, because each leaves remainder 1.

Therefore, *no* prime divides $g(y)$, for any $y \in \mathbb{Z}$, and hence the only possible values of the integers $g(y)$ are ± 1. In other words,

$$(p_1 p_2 \cdots p_k y)^2 + 1 = \pm 1 \quad \text{for all } y \in \mathbb{Z}.$$

But this is absurd, because each of the quadratic equations

$$(p_1 p_2 \cdots p_k y)^2 + 1 = 1 \quad \text{and} \quad (p_1 p_2 \cdots p_k y)^2 + 1 = -1$$

has at most two solutions y. This contradiction shows that $x^2 + 1$ is divisible by infinitely many primes, as required.   □

It now follows, by Fermat's two square theorem, that infinitely many primes are sums $a^2 + b^2$ of two squares. Hence there are infinitely many Gaussian primes $a + ib$ that are neither real nor pure imaginary.

Exercises

The argument just used to prove that $x^2 + 1$ is divisible by infinitely many primes can be generalized to any nonconstant polynomial $f(x)$ with integer coefficients. We suppose that

$$f(x) = a_m x^m + \cdots + a_1 x + a_0, \quad \text{where } a_0, a_1, \ldots, a_m \in \mathbb{Z} \text{ and } a_0, a_m \neq 0,$$

has values divisible only by the primes $p_1, p_2, \ldots, p_k$, and consider the polynomial

$$f(a_0 p_1 p_2 \cdots p_k y) = a_0 g(y),$$

where $g(y)$ is a polynomial of degree m.

6.7.1 Show that $g(y)$ has integer coefficients, constant term 1, and that any prime dividing a value of $g(y)$, for $y \in \mathbb{Z}$, also divides a value of $f(x)$, for $x \in \mathbb{Z}$.

6.7.2 Show, however, that none of $p_1, p_2 \ldots, p_k$ divide $g(y)$ when $y \in \mathbb{Z}$.

6.7.3 Deduce from Exercise 6.7.2 that $g(y) = \pm 1$ for any $y \in \mathbb{Z}$.

6.7.4 Show that the equations $g(y) = 1$ and $g(y) = -1$ have only finitely many solutions, which contradicts Exercise 6.7.3. (Where have you assumed that $f(x)$ is nonconstant?)

This contradiction shows that $f(x)$ is divisible by infinitely many primes. But now notice: we did *not* assume that infinitely many primes exist, hence this is a self-contained proof of Euclid's theorem that there are infinitely many primes.

6.7.5 Is this argument essentially different from Euclid's?

6.8   Discussion

The two square theorem was stated without proof by Fermat in 1640, though he claimed to have a proof by descent: assuming there is a prime that is of the form $4n + 1$ but *not* a sum of two squares he could show that there is a smaller prime with the same property. The first known proof of the theorem was in fact by descent, and published by Euler (1755). It cost him several years of effort.

Today it is possible to give quite simple proofs with the help of the result we called Lagrange's lemma in Section 6.5. Lagrange himself proved this lemma by means of Fermat's little theorem and his own theorem on the number of solutions of congruences mod p (Section 3.5).

Lagrange (1773) used his lemma together with his theory of equivalence of quadratic forms (Section 5.6) to give a new proof of the two square theorem. The part of the proof involving quadratic forms was simplified by Gauss (1801), long before his creation of the Gaussian integers. It seems that Gauss had the main results about $\mathbb{Z}[i]$, including unique prime factorization, around 1815, but they were first published in 1832. The proof used in this chapter, combining unique prime factorization in $\mathbb{Z}[i]$ with Lagrange's lemma, is due to Dedekind (1894).

Yet another popular proof uses the *geometry of numbers*, developed in the 1890s by Minkowski. It may be found in Scharlau and Opolka (1985) together with a historical introduction to Minkowski's results.

Parallel to all the popular proofs of the two square theorem there are analogous proofs of the *four square theorem* of Lagrange (1770): *every natural number is the sum of* (at most) four natural number squares. Most use the following counterpart of Lagrange's lemma: any prime p divides a number of the form $l^2 + m^2 + 1$. The counterpart turns out to be easier. What is harder is the *four square identity* discovered by Euler (1748b). It is analogous to the two square identity of Section 6.1 but is much more complicated (see Section 8.3). It can be mechanically checked by multiplying out both sides, but what does it mean?

The Gaussian integer proof is favored in this book because $\mathbb{Z}[i]$ is a natural structure and the two square identity is a natural part of it—the multiplicative property of the norm—rather than an accidental identity of formal expressions. In Chapter 8 we give a similar "structural" proof of the four square theorem that uses the *quaternion integers*. These are a remarkable four-dimensional structure from which the four square identity emerges naturally as the multiplicative property of the *quaternion norm*. Again, the key to the proof is unique prime factorization (or rather the prime divisor property, which happens to be somewhat easier than unique prime factorization in the quaternion case).

Fermat's two square theorem was generalized in another direction by Fermat himself. In 1654 Fermat announced similar theorems on primes of the forms $x^2 + 2y^2$ and $x^2 + 3y^2$:

$$p = x^2 + 2y^2 \Leftrightarrow p = 8n + 1 \text{ or } p = 8n + 3,$$
$$p = x^2 + 3y^2 \Leftrightarrow p = 3n + 1.$$

Our proof of the two square theorem in Section 6.5 can be adapted to Fermat's $x^2 + 2y^2$ and $x^2 + 3y^2$ theorems with the help of unique prime factorization theorems for numbers of the forms $a + b\sqrt{-2}$ and $a + b\sqrt{-3}$ respectively. Such theorems will be proved in the next chapter.

The other thing we need to adapt is Lagrange's lemma: if $p = 4n + 1$ then p divides a number of the form $m^2 + 1$ for some $m \in \mathbb{Z}$. In Section 6.7 we described this lemma (together with its converse) as the *quadratic character of* -1 because it says that -1 is congruent to a square mod p precisely when $p = 4n + 1$.

To prove Fermat's theorems on primes of the form $x^2 + 2y^2$ and $x^2 + 3y^2$ we similarly need the quadratic characters of -2 and -3. They are:

$$-2 \equiv \text{square (mod } p) \Leftrightarrow p = 8n + 1 \text{ or } 8n + 3,$$
$$-3 \equiv \text{square (mod } p) \Leftrightarrow p = 3n + 1.$$

Instead of finding quadratic characters one by one, in Chapter 9 we prove the sweeping *law of quadratic reciprocity*, which allows us to tell when any integer is congruent to a square mod p. Quadratic reciprocity was first observed by Euler and proved by him in special cases, such as those involved in Fermat's theorems. The first general proof is due to Gauss (1801), and quadratic reciprocity has since been proved in many different ways. In fact, it has been proved more often than any other theorem except its distant ancestor, the Pythagorean theorem.

7

Quadratic integers

PREVIEW

Just as Gaussian integers enable the factorization of $x^2 + y^2$, other quadratic expressions in ordinary integer variables are factorized with the help of *quadratic integers*. Examples in this chapter are

$$x^2 + 2 = (x - \sqrt{-2})(x + \sqrt{-2}),$$

$$x^2 - xy + y^2 = \left(x + \frac{-1 + \sqrt{-3}}{2} y\right)\left(x + \frac{-1 - \sqrt{-3}}{2} y\right).$$

In the first example, the factors belong to

$$\mathbb{Z}[\sqrt{-2}] = \{a + b\sqrt{-2} : a, b \in \mathbb{Z}\}.$$

Like the Gaussian integers $a + bi$, the numbers $a + b\sqrt{-2}$ enjoy unique prime factorization. We use this property to find all (ordinary) integer solutions of the equation $y^3 = x^2 + 2$.

The numbers $\frac{-1 + \sqrt{-3}}{2}$ and $\frac{-1 - \sqrt{-3}}{2}$ in the second example appear at first to be "fractional", and one might prefer to reserve the term "integer" for numbers in

$$\mathbb{Z}[\sqrt{-3}] = \{a + b\sqrt{-3} : a, b \in \mathbb{Z}\}.$$

However, unique prime factorization *fails* in $\mathbb{Z}[\sqrt{-3}]$, and it is precisely by adjoining the number $\frac{-1 + \sqrt{-3}}{2}$ that it is regained.

This leads to a discussion (partly in the exercises) of the general concept of quadratic integer and its applications. The chapter concludes with two remarkable applications of $\sqrt{-3}$: parametric formulas for the (infinitely many) rational solutions of $x^3 + y^3 = z^3 + w^3$, and proof that there are *no* nonzero integer solutions of $x^3 + y^3 = z^3$.

7.1   The equation $y^3 = x^2 + 2$

Diophantus was usually interested in rational solutions of equations, and the equation $y^3 = x^2 + 2$ has infinitely many of these (see Exercises 1.7.1–1.7.4). But in his *Arithmetica*, Book VI, Problem 17, Diophantus mentioned that $y^3 = x^2 + 2$ has the solution $x = 5, y = 3$. He evidently thought that this natural number solution was interesting. In 1657 Fermat added the claim that *there is no other natural number solution of $y^3 = x^2 + 2$*.

Fermat's claim was proved by Euler (1770), *assuming unique prime factorization in*

$$\mathbb{Z}[\sqrt{-2}] = \{a + b\sqrt{-2} : a, b \in \mathbb{Z}\}.$$

Euler gave no the proof of the latter fact (which is similar to the proof for $\mathbb{Z}[i]$ given in Section 6.4), but leaving unique factorization aside for the moment, his argument goes as follows.

Supposing that $y^3 = x^2 + 2$ for some $x, y \in \mathbb{Z}$, we have a factorization in $\mathbb{Z}[\sqrt{-2}]$:

$$y^3 = (x - \sqrt{-2})(x + \sqrt{-2}). \qquad (*)$$

Now we assume unique prime factorization in $\mathbb{Z}[\sqrt{-2}]$ and also that

$$\gcd(x - \sqrt{-2}, x + \sqrt{-2}) = 1$$

(another fact that we leave aside for the moment). By considering the prime factorizations of both sides of (*), it then follows that the factors $x - \sqrt{-2}$ and $x + \sqrt{-2}$ are cubes in $\mathbb{Z}[\sqrt{-2}]$.

The latter statement means that

$$
\begin{aligned}
x - \sqrt{-2} &= (a + b\sqrt{-2})^3 \quad \text{for some } a, b \in \mathbb{Z} \\
&= a^3 + 3a^2 b\sqrt{-2} - 6ab^2 - 2\sqrt{-2}b^3 \\
&= a^3 - 6ab^2 + (3a^2 b - 2b^3)\sqrt{-2}.
\end{aligned}
$$

Equating real and imaginary parts gives

$$
\begin{aligned}
x &= a^3 - 6ab^2 \\
1 &= 2b^3 - 3a^2 b = b(2b^2 - 3a^2).
\end{aligned}
$$

The latter equation says that the natural number b divides 1, hence $b = \pm 1$. Then the other divisor $2b^2 - 3a^2$ of 1 must equal -1, hence $a = \pm b = \pm 1$. This gives $x = \pm 5$ and hence $y = 3$. $\square$

Exercises

There is a similar treatment (using $\mathbb{Z}[i]$) of the equation $y^3 = x^2 + 1$, which shows that its only integer solution is $x = 0$, $y = 1$.

7.1.1 Supposing that the factors $x \pm i$ on the right-hand side of

$$y^3 = x^2 + 1 = (x - i)(x + i)$$

are cubes $(a \pm bi)^3$ in $\mathbb{Z}[i]$, deduce that $1 = b(3a^2 - b^2)$.

7.1.2 Deduce from Question 1 that $a = 0$, hence $x = 0$, hence $y = 1$.

A more challenging equation, which can also be mastered with the help of $\mathbb{Z}[i]$, is $y^3 = x^2 + 4$. Fermat claimed that the only natural number solutions are $x = 2$, $y = 2$ and $x = 11$, $y = 5$. Euler (1770) solved this equation using $\mathbb{Z}[i]$ but again without proving unique prime factorization. The argument goes along the above lines when x is odd, in which case it happens to be correct to assume that the factors $x - 2i$ and $x + 2i$ are both cubes.

7.1.3 Assuming that $x \pm 2i$ are cubes $(a \pm bi)^3$ in $\mathbb{Z}[i]$, show that $2 = b(3a^2 - b^2)$ and deduce that the positive odd $x = 11$.

7.2 The division property in $\mathbb{Z}[\sqrt{-2}]$

Unique prime factorization follows the same way as in $\mathbb{Z}$ and $\mathbb{Z}[i]$: from the prime divisor property that follows from a Euclidean algorithm, made possible by a division property.

Division property for $\mathbb{Z}[\sqrt{-2}]$. *For any α, $\beta \neq 0$ in $\mathbb{Z}[\sqrt{-2}]$ there are μ, ρ in $\mathbb{Z}[\sqrt{-2}]$ with*

$$\alpha = \mu\beta + \rho \quad and \quad |\rho| < |\beta|.$$

Proof. To see why the division property holds in $\mathbb{Z}[\sqrt{-2}]$ we look at the multiples $\mu\beta$ of any nonzero $\beta \in \mathbb{Z}[\sqrt{-2}]$. These lie at the corners of a grid of rectangles, the first of which has corners at 0, β, $\beta\sqrt{-2}$ and $\beta(1 + \sqrt{-2})$, shown in Figure 7.1.

Any $\alpha \in \mathbb{Z}[\sqrt{-2}]$ is in one of these rectangles and, as the picture shows, the distance $|\rho|$ of α from the nearest multiple $\mu\beta$ of β satisfies

$$|\rho|^2 \leq \left(\frac{|\beta|}{2}\right)^2 + \left(\frac{|\beta|}{\sqrt{2}}\right)^2 \quad \text{by Pythagoras' theorem}$$

$$= \frac{|\beta|^2 + 2|\beta|^2}{4} = \frac{3|\beta|^2}{4}.$$

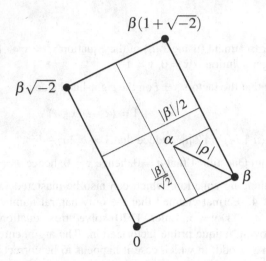

Figure 7.1: The division property in $\mathbb{Z}[\sqrt{-2}]$

Hence $|\rho| < |\beta|$, as required.   □

The units in $\mathbb{Z}[\sqrt{-2}]$, as in $\mathbb{Z}$, are just ± 1. We prove this using the *norm* $a^2 + 2b^2$ of $a + b\sqrt{-2}$. Units are elements of norm 1, and $a^2 + 2b^2 = 1$ only if $b = 0$ and $a = \pm 1$.

Now suppose we have a factorization of a cube into relatively prime factors s and t in $\mathbb{Z}[\sqrt{-2}]$:

$$y^3 = st.$$

Since s and t have no prime factor in common, the cubed prime factors of y^3 must separate into cubes inside s and cubes inside t. There could also be unit factors in s and t, but these can only be 1 or -1, *both of which are cubes*. Hence *the relatively prime factors of a cube are themselves cubes*.

This fills another gap in Euler's solution of $y^3 = x^2 + 2$. The only gap that now remains is to show that $\gcd(x - \sqrt{-2}, x + \sqrt{-2}) = 1$.

Exercises

The equation $y^3 = x^2 + 1$ that we took up in the last exercise set calls for a similar study of units in $\mathbb{Z}[i]$. Recall from Section 6.4 that the units of $\mathbb{Z}[i]$ are ± 1, $\pm i$.

7.2.1 Check that each of the units in $\mathbb{Z}[i]$ is a cube.

7.2.2 Deduce from Exercise 7.2.1 and unique prime factorization that relatively prime factors of a cube in $\mathbb{Z}[i]$ are themselves cubes.

The same properties of units and cubes apply to the equation $y^3 = x^2 + 4$ when we factorize its right-hand side into $(x - 2i)(x + 2i)$, but when x is even, say $x = 2X$, we have another problem.

7.2.3 Show that in this case y is even, $y = 2Y$ say, and hence $y^3 = x^2 + 4$ is equivalent to $2Y^3 = X^2 + 1 = (X - i)(X + i)$.

7.2.4 Show that X is odd in any integer solution of $2Y^3 = X^2 + 1$, and that in this case $1 - i$ divides $X - i$.

It follows, by taking conjugates, that $1 + i$ divides $X + i$ for any odd X. In fact, since $i(1 - i) = 1 + i$, the number $1 - i$ is a *common* divisor of $X - i$ and $X + i$ in $\mathbb{Z}[i]$. In the next exercise set we see whether $1 - i$ is their gcd.

7.3 The gcd in $\mathbb{Z}[\sqrt{-2}]$

Again we use the $\mathbb{Z}[\sqrt{-2}]$ norm

$$\text{norm}(a + b\sqrt{-2}) = |a + b\sqrt{-2}|^2 = a^2 + 2b^2,$$

which is multiplicative by the multiplicative property of absolute value.

Thus it is true, as in $\mathbb{Z}[i]$, that if α divides γ then $\text{norm}(\alpha)$ divides $\text{norm}(\gamma)$. Therefore, if δ is a common divisor of α and β, then $\text{norm}(\delta)$ is a common divisor of $\text{norm}(\alpha)$ and $\text{norm}(\beta)$.

We can now return to the equation

$$y^3 = x^2 + 2 = (x - \sqrt{-2})(x + \sqrt{-2})$$

and compute $\gcd(x - \sqrt{-2}, x + \sqrt{-2})$.

Relative primality of the factors. *If $x, y \in \mathbb{Z}$ are such that $y^3 = x^2 + 2$, then $\gcd(x - y\sqrt{-2}, x + y\sqrt{-2}) = 1$.*

Proof. If $y^3 = x^2 + 2$, then x must be odd. Indeed, for even x we have

$$x^2 + 2 \equiv 2 \pmod 4,$$

whereas

$$y^3 \equiv 0, 1 \text{ or } 3 \pmod 4.$$

This can be seen by trying $y \equiv 0, 1, 2, 3 \pmod 4$. It follows that the norm $x^2 + 2$ of $x \pm \sqrt{-2}$ is odd.

Now the gcd of $x - \sqrt{-2}$ and $x + \sqrt{-2}$ also divides their difference, $2\sqrt{-2}$, which has norm 8. The gcd of 8 and the odd number $x^2 + 2$ is 1. Therefore

$$\gcd(x - \sqrt{-2}, x + \sqrt{-2}) = 1. \qquad \qquad \square$$

We have now filled the gaps in Euler's proof that $x = 5$, $y = 3$ is the only natural number solution of $y^3 = x^2 + 2$: the cube y^3 factorizes into the relatively prime numbers $x - \sqrt{-2}$ and $x + \sqrt{-2}$, which are cubes by unique prime factorization in $\mathbb{Z}[\sqrt{-2}]$ and the fact that the units in $\mathbb{Z}[\sqrt{-2}]$ are also cubes. We can therefore write $x - \sqrt{-2} = (a + b\sqrt{-2})^3$ and complete the proof as already indicated in Section 7.1.

Exercises.

We can similarly use $\mathbb{Z}[i]$ to complete the proof, begun in the previous exercise sets, that $x = 0$, $y = 1$ is the only integer solution of $y^3 = x^2 + 1$.

7.3.1 Use congruence mod 4 to show that x is even in any integer solution of $y^3 = x^2 + 1$. From now on assume that (x, y) is such a solution.

7.3.2 Explain why $\gcd(x - i, x + i) = \gcd(x + i, 2)$ and use Question 7.3.1 to show that $\mathrm{norm}(x + i)$ is odd.

7.3.3 Deduce from Question 7.3.2 that $\gcd(x - i, x + i) = 1$.

7.3.4 Deduce, from the previous exercises and unique prime factorization in $\mathbb{Z}[i]$, that the factors on the right-hand side of $y^3 = (x - i)(x + i)$ are cubes in $\mathbb{Z}[i]$.

Likewise, we can find $\gcd(X - i, X + i)$ when X is odd, and hence complete the solution of $y^3 = x^2 + 4$ when $x = 2X$.

7.3.5 Show that 2 does not divide $X - i$ or $X + i$, and deduce from Exercise 7.2.4 that $\gcd(X - i, X + i) = 1 - i$.

7.3.6 Use unique prime factorization in $\mathbb{Z}[i]$ to deduce from $2Y^3 = (X - i)(X + i)$ that

$$X - i = (1 - i)(a - bi)^3 \quad \text{for some } a, b \in \mathbb{Z}.$$

7.3.7 Deduce from Exercise 7.3.6 that

$$1 = a^3 - b^3 + 3ab(a - b) = (a - b)(a^2 + 4ab + b^2)$$

and conclude that $X = 1$, hence $x = 2$.

7.4 $\mathbb{Z}[\sqrt{-3}]$ and $\mathbb{Z}[\zeta_3]$

A natural step after investigating $\mathbb{Z}[i]$ and $\mathbb{Z}[\sqrt{-2}]$ would be to study

$$\mathbb{Z}[\sqrt{-3}] = \{a + b\sqrt{-3} : a, b \in \mathbb{Z}\}.$$

But here is a surprise: *unique prime factorization fails in $\mathbb{Z}[\sqrt{-3}]$.*
    Consider the following factorizations of 4:

$$4 = 2 \times 2 = (1 - \sqrt{-3})(1 + \sqrt{-3}).$$

In $\mathbb{Z}[\sqrt{-3}]$ the norm is

$$\text{norm}(a + b\sqrt{-3}) = |a + b\sqrt{-3}|^2 = a^2 + 3b^2$$

and, as usual, if α divides γ then $\text{norm}(\alpha)$ divides $\text{norm}(\gamma)$.
    But now $\text{norm}(2) = 4$, which is not divisible by any smaller integer of the form $a^2 + 3b^2$ except 1, hence 2 is a prime of $\mathbb{Z}[\sqrt{-2}]$. And

$$\text{norm}(1 - \sqrt{-3}) = 1 + 3 = 4,$$

so $1 - \sqrt{-3}$ is also prime, as is $1 + \sqrt{-3}$. Thus 4 has two distinct prime factorizations in $\mathbb{Z}[\sqrt{-3}]$. $\square$

    This defect can be repaired by enlarging $\mathbb{Z}[\sqrt{-3}]$ to

$$\mathbb{Z}[\zeta_3] = \{a + b\zeta_3 : a, b \in \mathbb{Z}\},$$

where

$$\zeta_3 = \frac{-1 + \sqrt{-3}}{2}$$

is one of the imaginary cube roots of 1. (This is why we use the subscript 3. In general, ζ_n denotes $\cos\frac{2\pi}{n} + i\sin\frac{2\pi}{n}$, the nth root of 1.) The elements of $\mathbb{Z}[\zeta_3]$ lie at the corners of a tiling of the plane by equilateral triangles, and are called the *Eisenstein integers*.
    By geometric arguments like those used for $\mathbb{Z}[i]$ and $\mathbb{Z}[\sqrt{-2}]$, we can see that $\mathbb{Z}[\zeta_3]$ has the division property, hence a Euclidean algorithm and unique prime factorization. Figure 7.2 compares the point sets $\mathbb{Z}[\sqrt{-3}]$ and $\mathbb{Z}[\zeta_3]$ in the plane, showing why the division property fails for the first and succeeds for the second.
    In the rectangles of $\mathbb{Z}[\sqrt{-3}]$, each center point (such as the one at the top of the triangle shown) is at distance 1 from the two nearest corners,

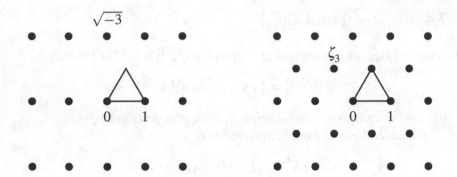

Figure 7.2: $\mathbb{Z}[\sqrt{-3}]$ (left) and $\mathbb{Z}[\zeta_3]$ (right)

hence its distance from the nearest corner is *not* less than the smallest side length of the rectangle. $\mathbb{Z}[\zeta_3]$ fills these "holes" in $\mathbb{Z}[\sqrt{-3}]$, producing the tiling of the plane by equilateral triangles.

Division property for $\mathbb{Z}[\zeta_3]$. *For any $\alpha, \beta \neq 0$ in $\mathbb{Z}[\zeta_3]$ there are μ, ρ in $\mathbb{Z}[\zeta_3]$ with*

$$\alpha = \mu\beta + \rho \quad \text{and} \quad |\rho| < |\beta|.$$

Proof. In the equilateral tiling, each point of the plane lies in some triangle and its distance from the nearest vertex is less than the side length of the triangle. In fact, its distance from *any* vertex of the surrounding triangle is less than the side length. This is so because a circle centered on a vertex and with a side as radius encloses the whole triangle.

This geometric property is the essence of the division property because, as usual, the set $\mu\beta$ of multiples of any nonzero $\beta \in \mathbb{Z}[\zeta_3]$ is the same shape as $\mathbb{Z}[\zeta_3]$. Its points are at the vertices of a tiling by equilateral triangles of side length $|\beta|$. Hence the distance $|\rho| = |\alpha - \mu\beta|$ from any $\alpha \in \mathbb{Z}[\zeta_3]$ to the nearest $\mu\beta$ is less than the side length $|\beta|$.   □

There are six units in $\mathbb{Z}[\zeta_3]$: ± 1, $\pm \zeta_3$ and $\pm \zeta_3^2$, and they lie at vertices of a hexagon on the unit circle with center at O (see Figure 7.2 again). Like the units in $\mathbb{Z}$ and $\mathbb{Z}[i]$, they all divide 1. The two distinct factorizations of 4 in $\mathbb{Z}[\sqrt{-3}]$ are the "same" in $\mathbb{Z}[\zeta_3]$, up to unit factors, thanks to the extra units in the latter (exercise).

Quadratic integers

It is satisfying to be able to repair the failure of unique prime factorization in $\mathbb{Z}[\sqrt{-3}]$ by extending it to $\mathbb{Z}[\zeta_3]$. But is it reasonable to consider $\zeta_3 = \frac{-1+\sqrt{-3}}{2}$ an "integer"? The general definition, which allows us to say yes, is the following.

Definition. A number $\alpha \in \mathbb{C}$ is an *algebraic integer* if it satisfies a *monic* polynomial equation with integer coefficients, that is

$$\alpha^m + a_{m-1}\alpha^{m-1} + \cdots + a_1\alpha + a_0 = 0 \quad \text{where} \quad a_0, a_1, \ldots, a_{m-1} \in \mathbb{Z}.$$

A *quadratic integer* is an algebraic integer satisfying a monic quadratic equation with integer coefficients.

We study the general concept of algebraic integer in Chapter 10, where it is shown that the sum, difference and product of algebraic integers are again algebraic integers. ζ_3 is an algebraic integer because it satisfies $x^3 - 1 = 0$. In fact, ζ_3 is a quadratic integer because it satisfies the equation $x^2 + x + 1 = 0$ obtained by factorizing $x^3 - 1$. All elements of $\mathbb{Z}[\zeta_3]$, being obtained from the algebraic integers 1 and ζ_3 by sum and difference, are algebraic integers. It can be shown directly that they are quadratic (exercise).

Closure under $+$, $-$, and $\times$ is certainly a natural requirement for integers, but perhaps the definition of algebraic integer goes too far and admits numbers that should not be considered integers. One reason that it does not is the following: *every rational algebraic integer is an ordinary integer*. This is crucial when results about ordinary integers are being derived as special cases of results about algebraic integers.

Rational algebraic integers. *If a rational number r satisfies a monic polynomial equation with integer coefficients, then r is an ordinary integer.*

Proof. Suppose that $r = s/t$, where $s, t \in \mathbb{Z}$ and $\gcd(s, t) = 1$, and suppose that r satisfies the equation

$$x^m + a_{m-1}x^{m-1} + \cdots + a_1x + a_0 = 0, \quad \text{where} \quad a_0, a_1, \ldots a_{m-1} \in \mathbb{Z}.$$

Substituting s/t for x and taking all terms but x^m to the right-hand side we get

$$\frac{s^m}{t^m} = -a_{m-1}\frac{s^{m-1}}{t^{m-1}} - \cdots - a_1\frac{s}{t} - a_0.$$

Multiplying both sides by t^m gives

$$s^m = -a_{m-1}s^{m-1}t - \cdots - a_1 st^{m-1} - a_0 t^m$$

$$= (-a_{m-1}s^{m-1} - \cdots - a_1 st^{m-2} - a_0 t^{m-1})t. \qquad (*)$$

Since $\gcd(s,t) = 1$, any prime factor of t divides the right-hand side but not the left. Hence $t = \pm 1$ by unique prime factorization in $\mathbb{Z}$, and therefore r is an ordinary integer.   □

Remark. It follows from this result that *the roots of a monic polynomial equation with integer coefficients are either integers or irrationals.* This generalizes what was proved in Section 5.1—that $\sqrt{n}$ is irrational when n is a nonsquare natural number—because $\sqrt{n}$ is a root of the equation $x^2 - n = 0$.

Exercises

The two factorizations of 4 found in $\mathbb{Z}[\sqrt{-3}]$, 2×2 and $(1 - \sqrt{-3})(1 + \sqrt{-3})$, differ only by unit factors in $\mathbb{Z}[\zeta_3]$.

7.4.1 Find units u and $\overline{u}$ in $\mathbb{Z}[\zeta_3]$ such that $2u \times 2\overline{u} = (1 - \sqrt{-3})(1 + \sqrt{-3})$.

7.4.2 Check that the units $\pm\zeta_3$, $\pm\zeta_3^2$ of $\mathbb{Z}[\zeta_3]$ satisfy monic polynomial equations with integer coefficients.

7.5   *Rational solutions of $x^3 + y^3 = z^3 + w^3$

Like the Pythagorean equation $x^2 + y^2 = z^2$, the equation $x^3 + y^3 = z^3 + w^3$ has infinitely many integer solutions, some of them of great renown. Here is one, associated with the great Indian number theorist Ramanujan.

> It was Littlewood who said that every positive integer was one of Ramanujan's personal friends. I remember going to see him once when he was lying ill at Putney. I had ridden in taxi-cab number 1729, and remarked that the number seemed to me rather a dull one, and that I hoped it was not an unfavorable omen. "No," he replied, "it is a very interesting number; it is the smallest number expressible as the sum of two cubes in two different ways."
>
> Hardy (1937).

The "two different ways" referred to by Ramanujan are

$$1729 = 9^3 + 10^3 = 1^3 + 12^3,$$

thus they correspond to an integer solution of $x^3 + y^3 = z^3 + w^3$. According to Brouncker (1657), the same solution was found by Frenicle, along with

$$9^3 + 15^3 = 2^3 + 16^3,$$
$$15^3 + 33^3 = 2^3 + 34^3,$$
$$16^3 + 33^3 = 9^3 + 34^3,$$
$$19^3 + 24^3 = 10^3 + 27^3.$$

Another startling solution of this equation is $x = 3$, $y = 4$, $z = -5$, $w = 6$; in other words $3^3 + 4^3 + 5^3 = 6^3$. This result, which seems to "generalize" $3^2 + 4^2 = 5^2$, really belongs to the equation $x^3 + y^3 = z^3 + w^3$, but the latter resembles the Pythagorean equation in one respect—there is a parametric formula for all its rational solutions.

The formula is due to Euler (1756). His method can be simplified using complex numbers, namely the norm in $\mathbb{Q}[\sqrt{-3}] = \{a + b\sqrt{-3} : a, b \in \mathbb{Q}\}$.

Parametric solution of $x^3 + y^3 = z^3 + w^3$. *The rational solutions are*

$$x = [(p + 3q)(p^2 + 3q^2) - 1]r,$$
$$y = [(-p + 3q)(p^2 + 3q^2) + 1]r,$$
$$z = [-p + 3q + (p^2 + 3q^2)^2]r,$$
$$w = [p + 3q - (p^2 + 3q^2)^2]r,$$

where p, q and r run through all rational numbers.

Proof. If we make the substitutions $x = X + Y$, $y = X - Y$, $z = Z + W$, $w = Z - W$, then the equation $x^3 + y^3 = z^3 + w^3$ becomes

$$X(X^2 + 3Y^2) = Z(Z^2 + 3W^2),$$

and X, Y, Z, W are rational if and only if x, y, z, w are.

Thus the problem is equivalent to finding the rational solutions of the equation $X(X^2 + 3Y^2) = Z(Z^2 + 3W^2)$. Also, we can specialize to $Z = 1$ (if we later multiply the solution by an arbitrary rational), so it suffices to find the rational solutions of

$$X = \frac{1 + 3W^2}{X^2 + 3Y^2}.$$

Now $a^2 + 3b^2 = |a + b\sqrt{-3}|^2$. Hence

$$X = \frac{|1 + W\sqrt{-3}|^2}{|X + Y\sqrt{-3}|^2}$$

$$= \left|\frac{1 + W\sqrt{-3}}{X + Y\sqrt{-3}}\right|^2 \qquad \text{since the norm is multiplicative}$$

$$= |p + q\sqrt{-3}|^2 = p^2 + 3q^2,$$

for some p, q that are rational if X, Y, W are.

We can define p and q as the real and imaginary parts of

$$p + q\sqrt{-3} = \frac{1 + W\sqrt{-3}}{X + Y\sqrt{-3}}.$$

Multiplying both sides by $X + Y\sqrt{-3}$ gives

$$pX - 3qY + (qX + pY)\sqrt{-3} = 1 + W\sqrt{-3},$$

and therefore, equating real and imaginary parts,

$$pX - 3qY = 1, \quad qX + pY = W.$$

Since these are linear equations in Y and W, we can solve for Y and W rationally in terms of p, q, and $X = p^2 + 3q^2$. This gives a 1-to-1 correspondence between rational pairs (p, q) and rational triples (X, Y, W) such that $X(X^2 + 3Y^2) = 1 + 3W^2$.

Substituting these values of X, Y, $Z = 1$, and W back in x, y, z, w we find that the rational solutions of $x^3 + y^3 = z^3 + w^3$ are all the rational multiples of

$$x = (p + 3q)(p^2 + 3q^2) - 1,$$
$$y = (-p + 3q)(p^2 + 3q^2) + 1,$$
$$z = -p + 3q + (p^2 + 3q^2)^2,$$
$$w = p + 3q - (p^2 + 3q^2)^2,$$

as claimed.   □

Example

$p = 1, q = 1$ gives

$$15^3 + 9^3 = 18^3 + (-12)^3,$$

which is a multiple (and rearrangement) of the equation $3^3 + 4^3 + 5^3 = 6^3$.

Exercises

7.5.1 Find a simpler p and q that also give (a multiple of) $3^3 + 4^3 + 5^3 = 6^3$.

It is not clear whether the parametric rational solution of $x^3 + y^3 = z^3 + w^3$ yields a parametric integer solution. However, Davenport (1960), p. 162, gives an infinite class of integer solutions discovered by Mahler in 1936.

7.5.2 By setting $p = 3q$ and then making a linear change of variable from q to t, derive the infinite family of integer solutions

$$x = 9t^3 - 1, \quad y = 1, \quad z = 9t^4, \quad w = 3t - 9t^4.$$

7.5.3 Find values of t that give $1^3 + 6^3 + 8^3 = 9^3$ and $9^3 + 10^3 = 1^3 + 12^3$.

7.6 *The prime $\sqrt{-3}$ in $\mathbb{Z}[\zeta_3]$

Perhaps the most important Diophantine equation that can be analyzed with the help of $\mathbb{Z}[\zeta_3]$ is the Fermat equation

$$x^3 + y^3 = z^3. \tag{*}$$

By doing so we settle the $n = 3$ case of *Fermat's last theorem*: $x^n + y^n \neq z^n$ for natural numbers x, y, z and $n > 2$.

Supposing, for the sake of contradiction, that (*) holds for some natural numbers x, y, z, we factorize the left-hand side:

$$x^3 + y^3 = (x+y)(x^2 - xy + y^2) \quad \text{in } \mathbb{Z}$$
$$= (x+y)(x + \zeta_3 y)(x + \zeta_3^2 y) \quad \text{in } \mathbb{Z}[\zeta_3].$$

If x, y, z are relatively prime, one might then hope that $x+y$, $x+\zeta_3 y$, $x+\zeta_3^2 y$ are also relatively prime. If so, we could use unique prime factorization in $\mathbb{Z}[\zeta_3]$ to conclude that $x+y$, $x+\zeta_3 y$, $x+\zeta_3^2 y$ are units times cubes and plan to derive a contradiction by this route.

Surprisingly, the very assumption that $x^3 + y^3 = z^3$ forces a factor $\sqrt{-3}$ into the equation. By suitable naming of terms, $\sqrt{-3}$ divides z and each of $x+y$, $x + \zeta_3 y$, and $x + \zeta_3^2 y$. This ruins the original plan but suggests a new one: *divide both sides of the equation (*) by $(\sqrt{-3})^3$ and build a new equation of the same form but "with fewer factors of $\sqrt{-3}$".* By slightly generalizing the Fermat equation, the new plan can be made to work. It leads to a contradiction by infinite descent because an integer equation in $\mathbb{Z}[\zeta_3]$ cannot be divided by the integer $\sqrt{-3}$ indefinitely.

To see how $\sqrt{-3}$ insinuates itself into the Fermat equation (*), we first develop a few of its basic properties. These involve *congruence* mod $\sqrt{-3}$, where, as usual,

$$\sigma \equiv \tau \pmod{\nu}$$

means that ν divides $\sigma - \tau$. In particular, $\sigma \equiv 0 \pmod{\sqrt{-3}}$ means that $\sqrt{-3}$ divides σ.

Figure 7.3 shows the congruence classes mod $\sqrt{-3}$ in $\mathbb{Z}[\zeta_3]$. There are three of them: the classes of 0 (black), 1 (white), and -1 (grey).

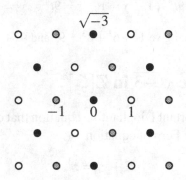

Figure 7.3: The congruence classes mod $\sqrt{-3}$

This can be checked by calculation. It suffices to look at the possible remainders on division by $\sqrt{-3}$, which have absolute value $< \sqrt{3}$. We can now prove the following properties:

Cubes mod 9. *For any $\sigma \in \mathbb{Z}[\zeta_3]$, if $\sigma \not\equiv 0 \pmod{\sqrt{-3}}$ then $\sigma^3 \equiv \pm 1$* (mod 9).

Proof. Since $\sigma \not\equiv 0$, $\sigma \equiv \pm 1 \pmod{\sqrt{-3}}$. Choose $\tau = \pm\sigma$ with $\tau \equiv 1$ (mod $\sqrt{-3}$), so $\tau = 1 + \mu\sqrt{-3}$ for some $\mu \in \mathbb{Z}[\zeta_3]$. Then

$$\begin{aligned}
\tau^3 - 1 &= (1 + \mu\sqrt{-3})^3 - 1 \\
&= 3\mu\sqrt{-3} + 3(\mu\sqrt{-3})^2 + (\mu\sqrt{-3})^3 \\
&= 3\sqrt{-3}(\mu + \mu^2\sqrt{-3} - \mu^3) \\
&\equiv 3\sqrt{-3}(\mu - \mu^3) \pmod{\sqrt{-3}} \\
&\equiv -3\sqrt{-3}\mu(\mu - 1)(\mu + 1) \pmod{\sqrt{-3}}.
\end{aligned}$$

Now μ, $\mu - 1$ and $\mu + 1$ are in different congruence classes, hence one of them is divisible by $\sqrt{-3}$. Thus $\tau^3 - 1$ is divisible by $-3\sqrt{-3}\sqrt{-3} = 9$. That is, $\tau^3 \equiv 1 \pmod 9$, and therefore $\sigma^3 \equiv \pm 1 \pmod 9$.   □

It follows from this property that if $\alpha, \beta, \gamma \in \mathbb{Z}[\zeta_3]$ and

$$\alpha^3 + \beta^3 + \gamma^3 = 0 \qquad\qquad (**)$$

—an equivalent of $x^3 + y^3 = z^3$ since $-z^3 = (-z)^3$—then $\sqrt{-3}$ divides at least one of α, β, γ. If not, then taking congruence mod 9 gives

$$\pm 1 \pm 1 \pm 1 \equiv 0 \pmod 9.$$

It easily checked that this is impossible for all eight combinations of signs. By suitable renaming of α, β, γ, if necessary, we can assume that γ is divisible by $\sqrt{-3}$, say $\gamma = \delta(\sqrt{-3})^n$. Then the three cubes equation (**) becomes

$$\alpha^3 + \beta^3 + \delta^3(\sqrt{-3})^{3n} = 0, \qquad\qquad (***)$$

and a second important property of congruence mod $\sqrt{-3}$ comes into play.

Congruence of factors in a sum of two cubes. *For any $\alpha, \beta \in \mathbb{Z}[\zeta_3]$, the factors $\alpha + \beta$, $\alpha + \zeta_3\beta$, $\alpha + \zeta_3^2\beta$ of $\alpha^3 + \beta^3$ are congruent mod $\sqrt{-3}$.*

Proof. Since $\zeta_3 = \frac{-1+\sqrt{-3}}{2}$,

$$\alpha + \zeta_3\beta = \alpha + \frac{-1+\sqrt{-3}}{2}\beta = \alpha + \beta + \frac{-3+\sqrt{-3}}{2}\beta$$

$$= \alpha + \beta + \frac{1+\sqrt{-3}}{2}\beta\sqrt{-3}$$

$$\equiv \alpha + \beta \pmod{\sqrt{-3}}$$

Similarly,

$$\alpha + \zeta_3^2\beta = \alpha + \frac{-1-\sqrt{-3}}{2}\beta \equiv \alpha + \beta \pmod{\sqrt{-3}} \qquad \square$$

We now apply this property to the factorized form of equation (***):

$$(\alpha + \beta)(\alpha + \zeta_3\beta)(\alpha + \zeta_3^2\beta) + \delta^3(\sqrt{-3})^{3n} = 0.$$

The number $\sqrt{-3}$ is prime in $\mathbb{Z}[\zeta_3]$ because its norm 3 is prime in $\mathbb{Z}$. Since $\sqrt{-3}$ divides $\delta^3(\sqrt{-3})^{3n}$, it also divides at least one of the factors $\alpha + \beta$, $\alpha + \zeta_3\beta$, $\alpha + \zeta_3^2\beta$. But then it divides all of them, since they are congruent mod $\sqrt{-3}$. Altogether we get: *if numbers $\alpha, \beta, \gamma \in \mathbb{Z}[\zeta_3]$ satisfy*

$$\alpha^3 + \beta^3 + \gamma^3 = 0,$$

then (with suitable renaming of numbers if necessary) $\sqrt{-3}$ divides γ and all the factors $\alpha + \beta$, $\alpha + \zeta_3\beta$, $\alpha + \zeta_3^2\beta$ in $\alpha^3 + \beta^3$.

Exercises

In other approaches to the equation $\alpha^3 + \beta^3 + \gamma^3 = 0$ that I know—Nagell (1951), p. 241, Grosswald (1966), p. 169, Redmond (1996), p. 697, and the outline in Baker (1984), p. 86—the number $\lambda = 1 - \zeta_3$ is used in place of $\sqrt{-3}$. This is probably because the equation $\alpha^n + \beta^n + \gamma^n = 0$ can be similarly treated using $\lambda = 1 - \zeta_n$, for certain other values of n, such as $n = 5$. However, it seems to me that $\sqrt{-3}$ is easier to use in the special case $n = 3$, though one can see with hindsight that $\lambda = 1 - \zeta_3$ does essentially the same job. The reason is the following.

7.6.1 Show that λ equals a unit times $\sqrt{-3}$.

7.6.2 Deduce from Exercise 7.6.1 that $\sigma \equiv \tau \pmod{\lambda} \Leftrightarrow \sigma \equiv \tau \pmod{\sqrt{-3}}$ and that $\sigma \equiv \tau \pmod{\lambda^4} \Leftrightarrow \sigma \equiv \tau \pmod 9$.

7.7   *Fermat's last theorem for $n = 3$

We now justify the hunch that the equation $\alpha^3 + \beta^3 + \gamma^3 = 0$ is impossible for $\alpha, \beta, \gamma \in \mathbb{Z}[\zeta_3]$ because it seemingly admits unlimited division by $\sqrt{-3}$. We suppose that γ is the term divisible by $\sqrt{-3}$, so $\gamma = (\sqrt{-3})^n \delta$ for some δ not divisible by $\sqrt{-3}$. Then the equation can be written

$$\alpha^3 + \beta^3 + (\sqrt{-3})^{3n} \delta^3 = 0$$

for some natural number n that we suppose to be as small as possible. In fact we must have $n \geq 2$. This is so because α, β, and γ are relatively prime, hence α and β, like δ, are not divisible by $\sqrt{-3}$. But then each is $\equiv \pm 1 \pmod{\sqrt{-3}}$ by the enumeration of congruence classes in the previous section. Hence if $n = 1$ and we reduce the equation mod 9, the property of cubes mod 9 gives

$$\pm 1 \pm 1 \pm (\sqrt{-3})^3 \equiv 0 \pmod 9$$

which is clearly impossible for any combination of signs.

We can therefore assume that $n \geq 2$. To enable repeated division by $\sqrt{-3}$ we assume that a slightly more general equation holds:

$$\alpha^3 + \beta^3 + \varepsilon(\sqrt{-3})^{3n} \delta^3 = 0, \tag{*}$$

where $\alpha, \beta, \delta \in \mathbb{Z}[\zeta_3]$ are relatively prime and ε is a unit of $\mathbb{Z}[\zeta_3]$. The unit ε is there because, as we shall see, division introduces units that cannot be completely eliminated.

Impossibility of $\alpha^3 + \beta^3 + \varepsilon(\sqrt{-3})^{3n}\delta^3 = 0$. *When* $\alpha, \beta, \delta \in \mathbb{Z}[\zeta_3]$ *are relatively prime and not divisible by* $\sqrt{-3}$, $n \geq 2$, *and* ε *is a unit,*

$$\alpha^3 + \beta^3 + \varepsilon(\sqrt{-3})^{3n}\delta^3 \neq 0.$$

Proof. Suppose on the contrary that there are relatively prime α, β, δ, not divisible by $\sqrt{-3}$, such that equation (*) holds. It follows from unique prime factorization in $\mathbb{Z}[\zeta_3]$ that the prime $\sqrt{-3}$ divides $\alpha^3 + \beta^3$, and hence $\sqrt{-3}$ divides at least one of its factors $\alpha + \beta$, $\alpha + \zeta_3\beta$, $\alpha + \zeta_3^2\beta$.

But all of these factors are congruent mod $\sqrt{-3}$, as we found in the last section, so in fact $\sqrt{-3}$ divides them all, and therefore

$$\frac{\alpha + \beta}{\sqrt{-3}}, \ \frac{\alpha + \zeta_3\beta}{\sqrt{-3}}, \ \frac{\alpha + \zeta_3^2\beta}{\sqrt{-3}} \in \mathbb{Z}[\zeta_3].$$

These three elements have no common prime divisor in $\mathbb{Z}[\zeta_3]$. For example, a common divisor of $\frac{\alpha + \beta}{\sqrt{-3}}$ and $\frac{\alpha + \zeta_3\beta}{\sqrt{-3}}$ also divides their difference,

$$\frac{1 - \zeta_3}{\sqrt{-3}}\beta = \frac{3 - \sqrt{-3}}{2\sqrt{-3}}\beta = \frac{1 + \sqrt{-3}}{2}\beta = \text{unit} \times \beta.$$

Hence a common divisor of $\frac{\alpha + \beta}{\sqrt{-3}}$ and $\frac{\alpha + \zeta_3\beta}{\sqrt{-3}}$ divides β. One finds that it also divides α by similarly considering $\frac{\alpha + \zeta_3\beta}{\sqrt{-3}} - \zeta_3\frac{\alpha + \beta}{\sqrt{-3}}$. But there is no common prime divisor of α and β, hence none of $\frac{\alpha + \beta}{\sqrt{-3}}$ and $\frac{\alpha + \zeta_3\beta}{\sqrt{-3}}$. Similar algebra shows the same for the other pairs from $\frac{\alpha + \beta}{\sqrt{-3}}$, $\frac{\alpha + \zeta_3\beta}{\sqrt{-3}}$, and $\frac{\alpha + \zeta_3^2\beta}{\sqrt{-3}}$.

Thus we can apply unique prime factorization to the following rearrangement of equation (*),

$$\frac{\alpha + \beta}{\sqrt{-3}} \cdot \frac{\alpha + \zeta_3\beta}{\sqrt{-3}} \cdot \frac{\alpha + \zeta_3^2\beta}{\sqrt{-3}} = -\varepsilon(\sqrt{-3})^{3n-3}\delta^3,$$

to conclude that each factor on the left is a unit times a cube, say

$$\frac{\alpha + \beta}{\sqrt{-3}} = \varepsilon_1\alpha_1^3, \quad \frac{\alpha + \zeta_3\beta}{\sqrt{-3}} = \varepsilon_2\beta_1^3, \quad \frac{\alpha + \zeta_3^2\beta}{\sqrt{-3}} = \varepsilon_3\gamma_1^3,$$

with α_1, β_1, γ_1 relatively prime because $\frac{\alpha + \beta}{\sqrt{-3}}$, $\frac{\alpha + \zeta_3\beta}{\sqrt{-3}}$, $\frac{\alpha + \zeta_3^2\beta}{\sqrt{-3}}$ are. It follows that the prime power $(\sqrt{-3})^{3n-3}$ resides in exactly one of α_1^3, β_1^3, γ_1^3. By renaming, if necessary, we can assume that it is in $\gamma_1^3 = (\sqrt{-3})^{3n-3}\delta_1^3$.

Now notice the delightful fact that

$$\zeta_3^2 \frac{\alpha+\beta}{\sqrt{-3}} + \frac{\alpha+\zeta_3\beta}{\sqrt{-3}} + \zeta_3 \frac{\alpha+\zeta_3^2\beta}{\sqrt{-3}} = 0 \quad \text{because} \quad 1+\zeta_3+\zeta_3^2 = 0.$$

In terms of α_1, β_1, δ_1, this fact is

$$\zeta_3^2\varepsilon_1\alpha_1^3 + \varepsilon_2\beta_1^3 + \zeta_3\varepsilon_3(\sqrt{-3})^{3n-3}\delta_1^3 = 0,$$

which, when divided by the unit $\zeta_3^2\varepsilon_1$, takes the form

$$\alpha_1^3 + \varepsilon_4\beta_1^3 + \varepsilon_5(\sqrt{-3})^{3n-3}\delta_1^3 = 0. \tag{**}$$

Here ε_4, ε_5 are units and α_1, β_1, δ_1 are relatively prime and not divisible by $\sqrt{-3}$. Equation (**) is of almost the same form as (*), except for the presence of the unit ε_4. Fortunately we can show that $\varepsilon_4 = \pm 1$ as follows.

Since $n \geq 2$, $(\sqrt{-3})^{3n-3}$ is divisible by $3\sqrt{-3}$ whereas $\alpha_1^3, \beta_1^3 \equiv \pm 1$ mod 9 (by the property of cubes mod 9) and hence also mod $3\sqrt{-3}$. Thus reducing (**) mod $3\sqrt{-3}$ gives

$$\pm 1 \pm \varepsilon_4 \equiv 0 \quad (\text{mod } 3\sqrt{-3}).$$

The only units satisfying this congruence are $\varepsilon_4 = \pm 1$, as required.

Equation (**) is therefore of the simpler form

$$\alpha_1^3 \pm \beta_1^3 + \varepsilon_5(\sqrt{-3})^{3n-3}\delta_1^3 = 0 \tag{***}$$

with $\alpha_1, \beta_1, \delta_1$ relatively prime and not divisible by $\sqrt{-3}$. Since $-\beta_1^3 = (-\beta_1)^3$, (***) is indeed of the same form as (*), *except that the exponent of $\sqrt{-3}$ is less.*

This contradicts the assumption that the exponent of $\sqrt{-3}$ in (*) is as small as possible, hence (*) does not hold. $\qquad\square$

Corollary. *The equation $x^3 + y^3 = z^3$ is impossible for integers $x, y, z \neq 0$.*

Proof. Suppose on the contrary that $x^3 + y^3 = z^3$ for integers $x, y, z \neq 0$. Dividing by any common divisor in $\mathbb{Z}[\zeta_3]$ we obtain an equation

$$\alpha^3 + \beta^3 + \gamma^3 = 0 \quad \text{with} \quad \alpha, \beta, \gamma \in \mathbb{Z}[\zeta_3] \text{ relatively prime.}$$

By suitably renaming the numbers, if necessary, we can assume that the multiple of $\sqrt{-3}$ is $\gamma = (\sqrt{-3})^n\delta$, where δ is not divisible by $\sqrt{-3}$. We then have an equation

$$\alpha^3 + \beta^3 + (\sqrt{-3})^{3n}\delta^3 = 0,$$

where $\alpha, \beta, \delta \in \mathbb{Z}[\zeta_3]$ are relatively prime and not divisible by $\sqrt{-3}$. This is a special case of the equation just proved to be impossible, therefore $x^3 + y^3 = z^3$ is impossible for integers $x, y, z \neq 0$. $\qquad\square$

Exercises

The impossibility of $\alpha^3 + \beta^3 + \gamma^3 = 0$ for nonzero $\alpha, \beta, \gamma \in \mathbb{Z}[\zeta_3]$ is probably the most difficult result in this book, so the reader may find the ideas in the proof a little slippery. The following exercises aim to provide a better grip by using the ideas again in a similar problem. They prove a theorem of Legendre that

$$\alpha^3 + \beta^3 + 3\gamma^3 = 0 \quad \text{is impossible for nonzero } \alpha, \beta, \gamma \in \mathbb{Z}[\zeta_3].$$

As above, the algebra introduces unknown units, so we need to show impossibility of the more general equation

$$\alpha^3 + \beta^3 + \varepsilon(\sqrt{-3})^{3n+2}\gamma^3 = 0 \tag{*}$$

where $\sqrt{-3}$ does not divide γ and ε is unit of $\mathbb{Z}[\zeta_3]$. The usual preliminary step of dividing by any common factors allows us to assume that α, β, γ have no common prime divisor and are not divisible by $\sqrt{-3}$. We also assume that the exponent of $\sqrt{-3}$ in (*) is as small as possible.

7.7.1 Explain why $\alpha^3 + \beta^3 + 3\gamma^3 = 0$ is a specialization of (*).

7.7.2 Reducing (*) mod 9 and using the property of cubes, show that $n \geq 1$ in (*).

7.7.3 Now use the congruence of factors of $\alpha^3 + \beta^3$, their relative primality, and unique prime factorization in $\mathbb{Z}[\zeta_3]$, to conclude from (*) that two of

$$\frac{\alpha + \beta}{\sqrt{-3}}, \frac{\alpha + \zeta_3\beta}{\sqrt{-3}}, \frac{\alpha + \zeta_3^2\beta}{\sqrt{-3}} \in \mathbb{Z}[\zeta_3]$$

are units times cubes and the third is a unit times a cube times $(\sqrt{-3})^{3n-1}$.

7.7.4 Deduce, from Exercise 7.7.3 and the "delightful fact" that there is a valid equation of the form

$$\varepsilon_1 \alpha_1^3 + \varepsilon_2 \beta_1^3 + \varepsilon_3 (\sqrt{-3})^{3n-1}\gamma_1^3 = 0,$$

or equivalently

$$\alpha_1^3 + \varepsilon_4 \beta_1^3 + \varepsilon_5 (\sqrt{-3})^{3n-1}\gamma_1^3 = 0, \tag{**}$$

where $\varepsilon_4, \varepsilon_5$ are units of $\mathbb{Z}[\zeta_3]$ and $\alpha_1, \beta_1, \gamma_1$ are not divisible by $\sqrt{-3}$.

7.7.5 By reducing (**) mod 3, show that $\varepsilon_4 = \pm 1$ (where is $n \geq 1$ used?). Deduce that (**) is equivalent to an equation of the form (*), but with a smaller power of $\sqrt{-3}$.

7.7.6 Conclude that equation (*) does not hold for any nonzero $\alpha, \beta, \gamma \in \mathbb{Z}[\zeta_3]$.

7.8   Discussion

In recent chapters we have seen many ways in which algebraic numbers illuminate the ordinary integers, and Diophantine equations in particular. At the simplest level, the *multiplicative norm* enables us to do such things as:

- Generate solutions of the Pell equation $x^2 - ny^2 = 1$ from the powers of $x_1 + y_1 \sqrt{n}$, where (x_1, y_1) is the smallest natural number solution.

- Find all rational solutions of $x^3 + y^3 = z^3 + w^3$.

At a more sophisticated level, certain rings of algebraic integers can be shown to have *unique prime factorization*, among them $\mathbb{Z}[i]$, $\mathbb{Z}[\sqrt{-2}]$ and $\mathbb{Z}[\zeta_3]$. This enables us to analyze algebraic factorizations such as

$$x^2 + y^2 = (x - yi)(x + yi)$$
$$x^3 + y^3 = (x + y)(x + \zeta_3 y)(x + \zeta_3^2 y)$$

and find solutions of certain equations in which they appear. For example:

- The primitive solutions of the Pythagorean equation $x^2 + y^2 = z^2$ can be found by factorizing $x^2 + y^2$ in $\mathbb{Z}[i]$.

- Fermat's theorem that each prime $p = 4n + 1$ is a sum of two squares can be proved by showing that p divides $m^2 + 1$, and factorizing $m^2 + 1$ in $\mathbb{Z}[i]$.

- The integer solutions of the Bachet equation $y^3 = x^2 + 2$ can be found by factorizing $x^2 + 2$ in $\mathbb{Z}[\sqrt{-2}]$.

- Nonexistence of of natural number solutions of $x^3 + y^3 = z^3$ can be proved by factorizing $x^3 + y^3$ in $\mathbb{Z}[\zeta_3]$.

But so far we have proved that unique prime factorization holds only in $\mathbb{Z}$, $\mathbb{Z}[i]$, $\mathbb{Z}[\sqrt{-2}]$, and $\mathbb{Z}[\zeta_3]$, and we have seen that it does *not* hold in $\mathbb{Z}[\sqrt{-3}]$. Therefore, there is no guarantee that we can push on with this approach to $\mathbb{Z}[\sqrt{-5}]$, $\mathbb{Z}[\sqrt{-6}]$, ... or to $\mathbb{Z}[\zeta_n]$ for higher values of n.

In Chapter 11 we show that unique prime factorization fails again in $\mathbb{Z}[\sqrt{-5}]$, and this time it cannot be repaired by filling obvious "holes" in $\mathbb{Z}[\sqrt{-5}]$, as we did in $\mathbb{Z}[\sqrt{-3}]$. The situation calls for some "ideal numbers" from mathematical outer space—it is not clear that they exist in $\mathbb{C}$ where the usual algebraic numbers come from.

This dire situation was first recognized by Kummer in the 1840s, and it came to light when Lamé (1847) published a faulty proof of Fermat's last theorem that $x^n + y^n \neq z^n$ for natural numbers x, y, z and $n > 2$. Lamé used the factorization

$$x^n + y^n = (x+y)(x+\zeta_n y) \cdots (x+\zeta_n^{n-1}y)$$

where $\zeta_n = \cos\frac{2\pi}{n} + i\sin\frac{2\pi}{n}$, as we did in Section 7.7 with $n = 3$. However, he assumed that $\mathbb{Z}[\zeta_n]$ has unique prime factorization, and Kummer showed that this is false for $n \geq 23$. Kummer was no doubt aware that it also fails in rings of quadratic integers such as $\mathbb{Z}[\sqrt{-5}]$, but he was more interested in $\mathbb{Z}[\zeta_n]$, called the *cyclotomic* ("circle-cutting") *integers* because $1, \zeta_n, \zeta_n^2, \ldots, \zeta_n^{n-1}$ cut the unit circle into n equal parts.

He introduced "ideal numbers" to restore unique prime factorization in $\mathbb{Z}[\zeta_n]$, and it enabled him to prove Fermat's last theorem for many values of n, though not all. More importantly, the "ideal" concept outgrew the cyclotomic integers and spread into algebra and algebraic geometry as well as number theory. The simpler examples like $\mathbb{Z}[\sqrt{-3}]$ and $\mathbb{Z}[\sqrt{-5}]$ were pointed out by Dedekind in the 1870s, in the course of giving a down-to-earth explanation of "ideal numbers". We follow Dedekind's approach in Chapter 11, and an English translation of Dedekind's own exposition may be found in Dedekind (1877).

It should be mentioned that Fermat's last theorem for $n = 4$ and $n = 7$ may be proved using only ordinary integers. A proof for $n = 4$ was given by Fermat himself—one proof in number theory that he actually wrote down—and variations of it appear in many books. Two variations may be found in Stillwell (1998), pp. 131–134. An elementary proof for $n = 7$ was discovered by V. A. Lebesgue (1840), and it was further simplified by Genocchi (1876), following the remarkable strategy of forming the sum of seventh powers of the roots of a cubic equation. This little-known proof may be found in Nagell (1951), pp. 248–251, and Ribenboim (1999), pp. 57–62.

8

The four square theorem

PREVIEW

In this chapter we prove that every natural number is the sum of four integer squares, following a proof of Hurwitz. This proof has been chosen because it resembles the proof of Fermat's two square theorem already given in Chapter 6, and because it introduces the *quaternions*, a mathematical structure with many beautiful algebraic and geometric features.

We define the quaternions to be the matrices $\begin{pmatrix} a+di & b+ci \\ -b+ci & a-di \end{pmatrix}$, where $a,b,c,d \in \mathbb{R}$, after verifying that the matrices $\begin{pmatrix} a & b \\ -b & a \end{pmatrix}$ behave like the complex numbers. In this representation, the norm is just the determinant, and its multiplicative property follows from the multiplicative property of determinants. On complex-number matrices, the determinant gives again the two square identity, and on quaternions it gives a *four square* identity.

"Quaternion integers" should be the quaternions with $a,b,c,d \in \mathbb{Z}$. However, these lack the division property. To bring it in we augment them with "half integer points" to form the so-called *Hurwitz integers*. We can then establish a Euclidean algorithm and a prime divisor property. (The quaternion product is noncommutative, which is a slight obstacle, but we get around it by taking care always to multiply and divide on the same side.)

The proof of the four square theorem then follows the proof of the two square theorem very closely.

- Using conjugates, any ordinary prime that is not a Hurwitz prime is shown to be a sum of four squares.

- If an ordinary prime p divides a Hurwitz integer product $\alpha\beta$, then p divides α or p divides β.

- Any ordinary odd prime p divides a natural number of the form $1 + l^2 + m^2$ (analogous to Lagrange's lemma in Section 6.5 but easier to prove).

- The number $1 + l^2 + m^2$ *factorizes* in the Hurwitz integers. Hence p is not a Hurwitz prime and therefore p is a sum of four squares.

- Since every natural number n is a product of odd primes and the prime 2 (which equals $0^2 + 0^2 + 1^2 + 1^2$), the four square identity shows that n is a sum of four squares.

8.1   Real matrices and $\mathbb{C}$

In this chapter we introduce 4-dimensional "hypercomplex numbers" called quaternions. A quaternion is easily defined as a 2×2 matrix of complex numbers, but to see why we might expect matrices to behave like numbers, we first show how to model the complex numbers by 2×2 *real* matrices.

For each $a + bi \in \mathbb{C}$, with real a and b, consider the matrix

$$M(a + bi) = \begin{pmatrix} a & b \\ -b & a \end{pmatrix}.$$

It is easy to check (exercise) that

$$\begin{aligned} M(a_1 + b_1 i) + M(a_2 + b_2 i) &= M(a_1 + a_2 + (b_1 + b_2)i) \\ &= M((a_1 + b_1 i) + (a_2 + b_2 i)), \end{aligned}$$

$$\begin{aligned} M(a_1 + b_1 i) M(a_2 + b_2 i) &= M(a_1 a_2 - b_1 b_2 + (a_1 b_2 + b_1 a_2)i) \\ &= M((a_1 + b_1 i)(a_2 + b_2 i)). \end{aligned}$$

Thus matrix sum and product correspond to complex sum and product, and therefore the matrices

$$\begin{pmatrix} a & b \\ -b & a \end{pmatrix} \quad \text{for} \quad a, b \in \mathbb{R}$$

behave exactly like the complex numbers $a + bi$.

Another way to see this is to write

$$\begin{pmatrix} a & b \\ -b & a \end{pmatrix} = a\begin{pmatrix} 1 & 0 \\ 0 & 1 \end{pmatrix} + b\begin{pmatrix} 0 & 1 \\ -1 & 0 \end{pmatrix} = a1 + bi.$$

The identity matrix

$$1 = \begin{pmatrix} 1 & 0 \\ 0 & 1 \end{pmatrix}$$

behaves like the number 1, and

$$i = \begin{pmatrix} 0 & 1 \\ -1 & 0 \end{pmatrix}$$

behaves like $\sqrt{-1}$. Indeed

$$i^2 = \begin{pmatrix} -1 & 0 \\ 0 & -1 \end{pmatrix} = -1.$$

Not only does this matrix representation of $\mathbb{C}$ have natural counterparts of 1 and i, it also has a natural interpretation of the norm on $\mathbb{C}$ as the *determinant*. This is so because

$$\text{norm}(a+bi) = a^2 + b^2 = \det\begin{pmatrix} a & b \\ -b & a \end{pmatrix}$$

The multiplicative property of the norm follows from the multiplicative property of the determinant:

$$\det\begin{pmatrix} a_1 & b_1 \\ -b_1 & a_1 \end{pmatrix} \det\begin{pmatrix} a_2 & b_2 \\ -b_2 & a_2 \end{pmatrix} = \det\left(\begin{pmatrix} a_1 & b_1 \\ -b_1 & a_1 \end{pmatrix}\begin{pmatrix} a_2 & b_2 \\ -b_2 & a_2 \end{pmatrix}\right).$$
$$(*)$$

And since the matrix product on the right-hand side equals

$$\begin{pmatrix} a_1a_2 - b_1b_2 & a_1b_2 + b_1a_2 \\ -a_1b_2 - b_1a_2 & a_1a_2 - b_1b_2 \end{pmatrix},$$

equation (*) gives a new way to derive the Diophantus two square identity. Replacing each $\det\begin{pmatrix} a & b \\ -b & a \end{pmatrix}$ in (*) by $a^2 + b^2$ we get

$$(a_1^2 + b_1^2)(a_2^2 + b_2^2) = (a_1a_1 - b_1b_2)^2 + (a_1b_2 + b_1a_2)^2.$$

A geometric property of multiplication

Here is a good place to point out a property of multiplication that we have previously observed in special cases in Chapters 6 and 7: multiplication of all members of $\mathbb{C}$ by some fixed nonzero $z_0 \in \mathbb{C}$ is a *similarity* or *shape-preserving map*, that is, it multiplies all distances by a constant (namely, $|z_0|$).

This is because the distance between complex numbers z_1 and z_2 equals $|z_2 - z_1|$. When multiplied by z_0, z_1 and z_2 are sent to $z_0 z_1$ and $z_0 z_2$, the distance between which is

$$|z_0 z_2 - z_0 z_1| = |z_0(z_2 - z_1)| = |z_0||z_2 - z_1|$$

by the multiplicative property of the norm.

We observed cases of this in Chapter 6, where multiplying $\mathbb{Z}[i]$ by some $\beta \neq 0$ gave a grid of the same square shape, and in Chapter 7 where multiplying $\mathbb{Z}[\sqrt{-2}]$ by $\beta \neq 0$ gave a grid of the same rectangular shape. In Section 8.4 we use the multiplicative property of the quaternion norm to show similarly that any nonzero multiple of the quaternion "integers" is a grid of the same shape in $\mathbb{R}^4$. (Here we use the word "grid" rather loosely, since the quaternion integers are not simply a grid of 4-dimensional cubes).

Exercises

8.1.1 Check that

$$M(a_1 + b_1 i) + M(a_2 + b_2 i) = M((a_1 + b_1 i) + (a_2 + b_2 i)),$$
$$M(a_1 + b_1 i)M(a_2 + b_2 i) = M((a_1 + b_1 i)(a_2 + b_2 i)).$$

Although multiplication by z_0 leaves the shape of any figure in the plane $\mathbb{C}$ unaltered, the figure may be rotated.

8.1.2 How is the amount of rotation related to z_0?

8.2 Complex matrices and $\mathbb{H}$

For each pair $\alpha, \beta \in \mathbb{C}$ we consider the matrix

$$\begin{pmatrix} \alpha & \beta \\ -\overline{\beta} & \overline{\alpha} \end{pmatrix},$$

which we call a *quaternion*.

The set of quaternions is called $\mathbb{H}$, after Hamilton, who discovered them in 1843 (the matrix definition, however, is due to Cayley (1858)).

It is easy to check that the sum and difference of quaternions are again quaternions. So, too, is the product because

$$\begin{pmatrix} \alpha_1 & \beta_1 \\ -\overline{\beta_1} & \overline{\alpha_1} \end{pmatrix} \begin{pmatrix} \alpha_2 & \beta_2 \\ -\overline{\beta_2} & \overline{\alpha_2} \end{pmatrix} = \begin{pmatrix} \alpha_3 & \beta_3 \\ -\overline{\beta_3} & \overline{\alpha_3} \end{pmatrix},$$

where

$$\alpha_3 = \alpha_1\alpha_2 - \beta_1\overline{\beta_2}, \quad \beta_3 = \alpha_1\beta_2 + \beta_1\overline{\alpha_2}.$$

This can be verified by matrix multiplication and complex conjugation.

The *norm* of a quaternion q is defined to be its determinant, hence if

$$q = \begin{pmatrix} \alpha & \beta \\ -\overline{\beta} & \overline{\alpha} \end{pmatrix} \text{ then norm}(q) \text{ is}$$

$$\det \begin{pmatrix} \alpha & \beta \\ -\overline{\beta} & \overline{\alpha} \end{pmatrix} = \alpha\overline{\alpha} + \beta\overline{\beta} = |\alpha|^2 + |\beta|^2.$$

The multiplicative property of determinants now gives a "complex two square identity" similar to the Diophantus two square identity:

$$(|\alpha_1|^2 + |\beta_1|^2)(|\alpha_2|^2 + |\beta_2|^2) = |\alpha_1\alpha_2 - \beta_1\overline{\beta_2}|^2 + |\alpha_1\beta_2 + \beta_1\overline{\alpha_2}|^2.$$

This identity was discovered by Gauss around 1820, but he left it un-published.

Remark on associativity

It is easy to find quaternions q_1 and q_2 such that $q_1q_2 \neq q_2q_1$ (exercise). In fact, they include the *quaternion units* discussed in the next section. However, quaternion multiplication is at least associative,

$$q_1(q_2q_3) = (q_1q_2)q_3,$$

since matrix multiplication is associative. This property can be checked laboriously by computing the matrices on both sides, but it is preferable to recall that each matrix represents a *function*, namely the linear map

$$\begin{pmatrix} x \\ y \end{pmatrix} \mapsto \begin{pmatrix} x \\ y \end{pmatrix} \begin{pmatrix} \alpha & \beta \\ \gamma & \delta \end{pmatrix},$$

and that matrix multiplication represents composition of functions.

Function composition is *always* associative, simply because $f_1(f_2 f_3)$ and $(f_1 f_2)f_3$ are the same function, since both send X to $f_1(f_2(f_3(X)))$.

Exercises

8.2.1 Verify that the product of two quaternions is the matrix claimed.

8.2.2 Find quaternions q_1 and q_2 such that $q_1 q_2 \neq q_2 q_1$.

The matrix representation of quaternions also shows that a nonzero quaternion has a *multiplicative inverse*, namely its matrix inverse.

8.2.3 Compute the inverse of a nonzero quaternion $q = \begin{pmatrix} \alpha & \beta \\ -\bar{\beta} & \bar{\alpha} \end{pmatrix}$ and verify that q^{-1} is also a quaternion.

The complex two square identity is one way to derive the four square identity, written down in the next section.

8.2.4 By writing

$$\alpha_1 = a_1 + d_1 i, \quad \beta_1 = b_1 + c_1 i, \quad \alpha_2 = a_2 + d_2 i, \quad \beta_2 = b_2 + c_2 i,$$

express $(a_1^2 + b_1^2 + c_1^2 + d_1^2)(a_2^2 + b_2^2 + c_2^2 + d_2^2)$ as a sum of four squares.

8.3 The quaternion units

If we write $\alpha = a + di$ and $\beta = b + ci$, where $a, b, c, d \in \mathbb{R}$, then each quaternion can be viewed as a linear combination of four special matrices **1, i, j, k** called *quaternion units*.

$$\begin{pmatrix} \alpha & \beta \\ -\bar{\beta} & \bar{\alpha} \end{pmatrix} = \begin{pmatrix} a + di & b + ci \\ -b + ci & a - di \end{pmatrix}$$

$$= a \begin{pmatrix} 1 & 0 \\ 0 & 1 \end{pmatrix} + b \begin{pmatrix} 0 & 1 \\ -1 & 0 \end{pmatrix} + c \begin{pmatrix} 0 & i \\ i & 0 \end{pmatrix} + d \begin{pmatrix} i & 0 \\ 0 & -i \end{pmatrix}$$

$$= a\mathbf{1} + b\mathbf{i} + c\mathbf{j} + d\mathbf{k}.$$

The matrices **1, i, j, k** are quaternions of norm 1 that satisfy the following easily verified relations:

$$\mathbf{i}^2 = \mathbf{j}^2 = \mathbf{k}^2 = -1,$$
$$\mathbf{ij} = \mathbf{k} = -\mathbf{ji},$$
$$\mathbf{jk} = \mathbf{i} = -\mathbf{kj},$$
$$\mathbf{ki} = \mathbf{j} = -\mathbf{ik}.$$

Thus the product of quaternions is generally *noncommutative*:

$$q_1 q_2 \neq q_2 q_1.$$

Apart from this, however, the quaternions have the same basic properties as numbers. They form an abelian group under addition, the nonzero quaternions form a group under multiplication, and we also have

$$q_1(q_2 + q_3) = q_1 q_2 + q_1 q_3,$$
$$(q_2 + q_3)q_1 = q_2 q_1 + q_3 q_1$$

(left and right distributive laws).

The four square identity

If $q = a1 + b\mathbf{i} + c\mathbf{j} + d\mathbf{k}$ then norm(q) is

$$\det \begin{pmatrix} a+di & b+ci \\ -b+ci & a-di \end{pmatrix} = a^2 + b^2 + c^2 + d^2.$$

Since $\det(q_1)\det(q_2) = \det(q_1 q_2)$, we can also write the "complex two square identity" as a *real four square identity*, which turns out to be

$$
\begin{aligned}
(a_1^2 + b_1^2 + c_1^2 + d_1^2)(a_2^2 + b_2^2 + c_2^2 + d_2^2) = \quad & (a_1 a_2 - b_1 b_2 - c_1 c_2 - d_1 d_2)^2 \\
& + (a_1 b_2 + b_1 a_2 + c_1 d_2 - d_1 c_2)^2 \\
& + (a_1 c_2 - b_1 d_2 + c_1 a_2 + d_1 b_2)^2 \\
& + (a_1 d_2 + b_1 c_2 - c_1 b_2 + d_1 a_2)^2.
\end{aligned}
$$

Remarkably, the four square identity was discovered by Euler in 1748, nearly 100 years before the discovery of quaternions. Euler hoped to use it to prove that every natural number is the sum of four squares, by proving also that every *prime* is the sum of four squares. This was first proved by Lagrange in 1770. We can now give a simpler proof with the help of quaternions. This will be done in the next few sections.

Exercises

As mentioned in the previous section, Hamilton did not introduce quaternions as particular 2×2 matrices. He defined them directly as abstract objects of the form $a1 + b\mathbf{i} + c\mathbf{j} + d\mathbf{k}$, with multiplication defined by the following rules:

$$\mathbf{i}^2 = \mathbf{j}^2 = \mathbf{k}^2 = \mathbf{ijk} = -1.$$

8.3.1 Deduce from these relations that $\mathbf{ij} = \mathbf{k}$. Where does your computation assume associativity?

One can similarly find the product of any two units, and then the product of any two quaternions.

8.3.2 Explain the role of the distributive laws in computing products.

The eight units and their negatives form an interesting finite group.

8.3.3 Show that $Q = \{\pm 1, \pm\mathbf{i}, \pm\mathbf{j}, \pm\mathbf{k}\}$ is closed under products and inverses, and hence forms a (nonabelian) group under the quaternion product.

8.3.4 Show that the products of any two of $\mathbf{i}$, $\mathbf{j}$, $\mathbf{k}$, or their negatives, make up all of Q.

8.3.5 Deduce from Exercise 8.3.4 that any *proper* subgroup of Q (that is, any subgroup that is not all of Q) is abelian.

Q is in fact the smallest nonabelian group whose proper subgroups are all abelian.

8.4 $\mathbb{Z}[\mathbf{i},\mathbf{j},\mathbf{k}]$

From now on we write the quaternion $\mathbf{1}$ simply as 1 and omit it altogether as a term in a product. Thus the typical quaternion will be written

$$q = a + b\mathbf{i} + c\mathbf{j} + d\mathbf{k}, \quad \text{where} \quad a,b,c,d \in \mathbb{R}.$$

Which of these objects should be regarded as "integers"?
   One's first thought is that

$$\mathbb{Z}[\mathbf{i},\mathbf{j},\mathbf{k}] = \{a + b\mathbf{i} + c\mathbf{j} + d\mathbf{k} : a,b,c,d \in \mathbb{Z}\}$$

should be the "quaternion integers", analogous to the Gaussian integers $\mathbb{Z}[i]$. Sum, difference and product of members of $\mathbb{Z}[\mathbf{i},\mathbf{j},\mathbf{k}]$ are again members of $\mathbb{Z}[\mathbf{i},\mathbf{j},\mathbf{k}]$, and

$$\text{norm}(a + b\mathbf{i} + c\mathbf{j} + d\mathbf{k}) = a^2 + b^2 + c^2 + d^2$$

is an ordinary integer, which we can use to find "primes" in $\mathbb{Z}[\mathbf{i},\mathbf{j},\mathbf{k}]$.

Example. $2 + \mathbf{i} + \mathbf{j} + \mathbf{k}$ is a prime of $\mathbb{Z}[\mathbf{i},\mathbf{j},\mathbf{k}]$.

   This is so because $\text{norm}(2 + \mathbf{i} + \mathbf{j} + \mathbf{k}) = 2^2 + 1^2 + 1^2 + 1^2 = 7$, which is a prime in $\mathbb{Z}$. Hence $2 + \mathbf{i} + \mathbf{j} + \mathbf{k}$ is not the product of members of $\mathbb{Z}[\mathbf{i},\mathbf{j},\mathbf{k}]$ with smaller norm.

However, there is trouble when we attempt division with remainder: the set of "integer multiples" of a fixed quaternion has the wrong shape. Even though multiples are not visualizable as they are in $\mathbb{Z}[i]$—since the quaternions

$$\mathbb{H} = \{a + b\mathbf{i} + c\mathbf{j} + d\mathbf{k} : a, b, c, d \in \mathbb{R}\}$$

form a 4-dimensional space $\mathbb{R}^4$—we can nevertheless talk about distance and angles in $\mathbb{R}^4$ and reason geometrically.

Multiples in $\mathbb{Z}[\mathbf{i}, \mathbf{j}, \mathbf{k}]$

We interpret 1, $\mathbf{i}$, $\mathbf{j}$, $\mathbf{k}$ as the unit points on four perpendicular axes in $\mathbb{R}^4$. Then the quaternion norm $a^2 + b^2 + c^2 + d^2$ is just the square of the distance $|a + b\mathbf{i} + c\mathbf{j} + d\mathbf{k}|$ of $a + b\mathbf{i} + c\mathbf{j} + d\mathbf{k}$ from O.

More generally, $\mathrm{norm}(q_1 - q_2)$ is the square of the distance $|q_1 - q_2|$ between the quaternions q_1 and q_2.

Now, since the norm is multiplicative, we have

$$|qq_1 - qq_2| = |q(q_1 - q_2)| = |q||q_1 - q_2|,$$

so *multiplying all of* $\mathbb{H} = \mathbb{R}^4$ *by a quaternion* q *multiplies all distances by the constant* $|q|$. (Since $q \cdot 0 = 0$, multiplication by q also leaves the origin fixed, so when $|q| = 1$ this operation can be regarded as a "rotation" of $\mathbb{R}^4$ about O.)

It follows, if $q \neq 0$, that multiplication by q leaves all angles unchanged. In particular, the multiples β, $\beta\mathbf{i}$, $\beta\mathbf{j}$, and $\beta\mathbf{k}$ of 1, $\mathbf{i}$, $\mathbf{j}$, $\mathbf{k}$ by a quaternion $\beta \neq 0$ are each at distance $|\beta|$ from O and in perpendicular directions, like 1, $\mathbf{i}$, $\mathbf{j}$, $\mathbf{k}$.

Any multiple of β by an element of $\mathbb{Z}[\mathbf{i}, \mathbf{j}, \mathbf{k}]$ is just a sum of elements $\pm\beta$, $\pm\beta\mathbf{i}$, $\pm\beta\mathbf{j}$ and $\pm\beta\mathbf{k}$. Hence the multiples of β lie at corners of a "grid" like the points of $\mathbb{Z}[\mathbf{i}, \mathbf{j}, \mathbf{k}]$ itself—a grid of what we call *4-cubes*. The only difference is that the grid of multiples of β is magnified by $|\beta|$ and possibly rotated.

Exercises

The rotations of $\mathbb{R}^4$ obtained by multiplying each point by a quaternion $q \neq 1$ with $|q| = 1$ are unlike rotations of $\mathbb{R}^3$ in that they have no "axis" of fixed points.

8.4.1 Show that multiplication by a quaternion $q \neq 1$ fixes only the origin.

The detection of quaternion primes by their norms, which are sums of squares, allows us to prove already that there are infinitely many of them.

8.4.2 Without assuming that every natural number is a sum of four squares, show that there are infinitely many quaternion primes.

8.5 The Hurwitz integers

Division with remainder

Just as in $\mathbb{Z}[i]$, so too in $\mathbb{Z}[\mathbf{i},\mathbf{j},\mathbf{k}]$ we look at the grid of multiples of β to find the remainder when α is divided by β. It is $\alpha - \mu\beta$, where $\mu\beta$ is the nearest corner of the grid.

Unfortunately, we do not always have $|\alpha - \mu\beta| < |\beta|$. There is one exceptional position: if α lies at the center of one of the 4-cubes, then $|\alpha - \mu\beta| = |\beta|$. This is because the center-to-corner distance in any 4-cube equals the length of a side. For example, the center

$$\frac{1}{2} + \frac{\mathbf{i}}{2} + \frac{\mathbf{j}}{2} + \frac{\mathbf{k}}{2}$$

of the unit 4-cube with edges on the axes has distance from O

$$\sqrt{\left(\frac{1}{2}\right)^2 + \left(\frac{1}{2}\right)^2 + \left(\frac{1}{2}\right)^2 + \left(\frac{1}{2}\right)^2} = \sqrt{1} = 1.$$

Thus *the division property fails for* $\mathbb{Z}[\mathbf{i},\mathbf{j},\mathbf{k}]$. In this respect, $\mathbb{Z}[\mathbf{i},\mathbf{j},\mathbf{k}]$ is like $\mathbb{Z}[\sqrt{-3}]$ rather than $\mathbb{Z}[i]$. Indeed, we fix the problem exactly as we did for $\mathbb{Z}[\sqrt{-3}]$, by adding the exceptional points as extra integers.

Since each midpoint is obtained by adding $\frac{1+\mathbf{i}+\mathbf{j}+\mathbf{k}}{2}$ to some member of $\mathbb{Z}[\mathbf{i},\mathbf{j},\mathbf{k}]$, we want the set of quaternions of the form

$$\frac{1+\mathbf{i}+\mathbf{j}+\mathbf{k}}{2} + a + b\mathbf{i} + c\mathbf{j} + d\mathbf{k} \quad \text{for} \quad a,b,c,d \in \mathbb{Z}$$

together with those in $\mathbb{Z}[\mathbf{i},\mathbf{j},\mathbf{k}]$,

$$a + b\mathbf{i} + c\mathbf{j} + d\mathbf{k} \quad \text{for} \quad a,b,c,d \in \mathbb{Z}.$$

A single formula that embraces both these sets of points is

$$A\frac{1+\mathbf{i}+\mathbf{j}+\mathbf{k}}{2} + B\mathbf{i} + C\mathbf{j} + D\mathbf{k} \quad \text{for} \quad A,B,C,D \in \mathbb{Z}.$$

We get the midpoints for A odd and the points of $\mathbb{Z}[\mathbf{i},\mathbf{j},\mathbf{k}]$ for A even. Thus the quaternions we have constructed to ensure the division property are the set $\mathbb{Z}[\frac{1+\mathbf{i}+\mathbf{j}+\mathbf{k}}{2},\mathbf{i},\mathbf{j},\mathbf{k}]$ of all the integer combinations of

$$\frac{1+\mathbf{i}+\mathbf{j}+\mathbf{k}}{2},\quad \mathbf{i},\quad \mathbf{j},\quad \mathbf{k}.$$

The quaternions in $\mathbb{Z}[\frac{1+\mathbf{i}+\mathbf{j}+\mathbf{k}}{2},\mathbf{i},\mathbf{j},\mathbf{k}]$ are called the *Hurwitz integers*, after Hurwitz, who introduced them in 1896. We are going to follow his idea of using them to prove that every natural number is the sum of four integer squares. (This approach may also be found in Hardy and Wright (1979) and in Samuel (1970).)

But first, why should these things be regarded as "integers"?

1. The sum and difference of Hurwitz integers are clearly Hurwitz integers.

2. It can be checked (with more difficulty) that the product of Hurwitz integers is a Hurwitz integer.

3. It can also be checked that the norm of a Hurwitz integer is an ordinary integer.

Example. $\frac{7+\mathbf{i}+\mathbf{j}+\mathbf{k}}{2} = \frac{1+\mathbf{i}+\mathbf{j}+\mathbf{k}}{2} + 3$

This Hurwitz integer has norm

$$\frac{7^2+1^2+1^2+1^2}{4} = \frac{52}{4} = 13.$$

Since 13 is an ordinary prime, $\frac{7+\mathbf{i}+\mathbf{j}+\mathbf{k}}{2}$ is not the product of Hurwitz integers of smaller norm, hence it is a *Hurwitz prime*.

Exercises

8.5.1 Write each of $1,\mathbf{i},\mathbf{j},\mathbf{k}$ in the form

$$A\frac{1+\mathbf{i}+\mathbf{j}+\mathbf{k}}{2} + B\mathbf{i} + C\mathbf{j} + D\mathbf{k} \quad \text{for} \quad A,B,C,D \in \mathbb{Z},$$

and thus show that $\mathbb{Z}[\frac{1+\mathbf{i}+\mathbf{j}+\mathbf{k}}{2},\mathbf{i},\mathbf{j},\mathbf{k}]$ includes $\mathbb{Z}[\mathbf{i},\mathbf{j},\mathbf{k}]$. Also show that the norm of each Hurwitz integer is an ordinary integer.

The units of $\mathbb{Z}[\frac{1+\mathbf{i}+\mathbf{j}+\mathbf{k}}{2},\mathbf{i},\mathbf{j},\mathbf{k}]$ are the eight units $\pm 1, \pm\mathbf{i}, \pm\mathbf{j}, \pm\mathbf{k}$ of $\mathbb{Z}[\mathbf{i},\mathbf{j},\mathbf{k}]$, together with the 16 midpoints $\pm\frac{1}{2} \pm \frac{\mathbf{i}}{2} \pm \frac{\mathbf{j}}{2} \pm \frac{\mathbf{k}}{2}$ nearest to the origin. Like the units of $\mathbb{Z}[i]$ or $\mathbb{Z}[\zeta_3]$, they form a group, since the set of them is closed under products and inverses. However, the group of units of the Hurwitz integers is much more interesting, because it is larger and also nonabelian.

8.5.2 Show that the 24 units listed above include the product of any two of them.

8.5.3 Deduce from the product calculations in Exercise 8.5.2 that the 24 units include the inverse of any one of them.

8.6 Conjugates

For any quaternion $q = a + b\mathbf{i} + c\mathbf{j} + d\mathbf{k}$ we call

$$\overline{q} = a - b\mathbf{i} - c\mathbf{j} - d\mathbf{k}$$

the *conjugate* of q. This conjugate has almost the same basic properties as the complex conjugate:

$$q\overline{q} = |q|^2,$$
$$\overline{q_1 + q_2} = \overline{q_1} + \overline{q_2},$$
$$\overline{q_1 - q_2} = \overline{q_1} - \overline{q_2},$$
$$\overline{q_1 q_2} = \overline{q_2}\,\overline{q_1}.$$

(The reversal of the product in the last one is due to noncommutative quaternion multiplication.)

As in $\mathbb{C}$, the properties of conjugation in $\mathbb{H}$ can be checked by working out both sides. We use them (much as we did in Section 6.3) to prove a conditional four square theorem: *if p is an ordinary prime but not a Hurwitz prime then*

$$p = a^2 + b^2 + c^2 + d^2 \quad where \quad 2a, 2b, 2c, 2d \in \mathbb{Z}.$$

Suppose p has a nontrivial Hurwitz integer factorization

$$p = (a + b\mathbf{i} + c\mathbf{j} + d\mathbf{k})\gamma.$$

Then, taking conjugates of both sides, we get

$$p = \overline{\gamma}(a - b\mathbf{i} - c\mathbf{j} - d\mathbf{k}), \quad \text{since } \overline{p} = p.$$

Multiplying the two expressions for p gives

$$p^2 = (a+b\mathbf{i}+c\mathbf{j}+d\mathbf{k})\gamma\bar{\gamma}(a-b\mathbf{i}-c\mathbf{j}-d\mathbf{k})$$
$$= (a+b\mathbf{i}+c\mathbf{j}+d\mathbf{k})(a-b\mathbf{i}-c\mathbf{j}-d\mathbf{k})\gamma\bar{\gamma} \quad \text{since } \gamma\bar{\gamma} \text{ is real}$$
$$= (a^2+b^2+c^2+d^2)|\gamma|^2,$$

where both $a^2+b^2+c^2+d^2$, $|\gamma|^2 > 1$. But the only positive integer factorization of p^2 is pp, hence $p = a^2+b^2+c^2+d^2$.

Finally, since a,b,c,d are the coefficients of a Hurwitz integer, they could be half integers, but at any rate $2a,2b,2c,2d \in \mathbb{Z}$. $\quad\square$

Varying the factors of p

By finding a new factorization of p, we now show that *any ordinary prime that is not a Hurwitz prime is the sum of four integer squares.*

A Hurwitz integer α with half-integer coordinates can always be written in the form

$$\alpha = \omega + a' + b'\mathbf{i} + c'\mathbf{j} + d'\mathbf{k},$$

where a',b',c',d' are *even* integers, by a suitable choice of signs in the Hurwitz integer

$$\omega = \frac{\pm 1 \pm \mathbf{i} \pm \mathbf{j} \pm \mathbf{k}}{2}.$$

The norm of ω is 1, so that $\omega\bar{\omega} = 1$.

Now suppose we have $p = a^2+b^2+c^2+d^2$ for an ordinary prime p, as in the last subsection, and that a,b,c,d are half integers. We first write

$$p = (a+b\mathbf{i}+c\mathbf{j}+d\mathbf{k})(a-b\mathbf{i}-c\mathbf{j}-d\mathbf{k})$$
$$= (\omega + a' + b'\mathbf{i} + c'\mathbf{j} + d'\mathbf{k}) \times (\bar{\omega} + a' - b'\mathbf{i} - c'\mathbf{j} - d'\mathbf{k})$$

where a',b',c',d' are even and ω is as above, so $\omega\bar{\omega} = 1$. Next we insert $1 = \bar{\omega}\omega$ between the (conjugate) factors just found, and in this way obtain new conjugate factors of p,

$$p = (\omega + a' + b'\mathbf{i} + c'\mathbf{j} + d'\mathbf{k})\bar{\omega} \times \omega(\bar{\omega} + a' - b'\mathbf{i} - c'\mathbf{j} - d'\mathbf{k}).$$

In the first factor, ω plus the even integer terms times $\bar{\omega}$ gives 1 plus integer terms, hence it is

$$A + B\mathbf{i} + C\mathbf{j} + D\mathbf{k} \quad \text{for some} \quad A,B,C,D \in \mathbb{Z}.$$

The second factor is its conjugate, hence

$$p = A^2 + B^2 + C^2 + D^2 \quad \text{with} \quad A,B,C,D, \in \mathbb{Z}. \quad\square$$

Exercises

The proof above shows that any sum of integer squares has a nontrivial quaternion integer factorization.

8.6.1 Find quaternion integer factorizations of the Gaussian primes 3, 7, and 11.

8.6.2 Why are these Gaussian primes?

The properties of conjugates enumerated above can be proved using matrices or using the multiplication rules for **i**, **j**, and **k**.

8.6.3 If $q = \begin{pmatrix} \alpha & \beta \\ -\overline{\beta} & \overline{\alpha} \end{pmatrix}$ what is $\overline{q}$?

8.6.4 Use the matrix just found for $\overline{q}$ to compute $\overline{q_2}\,\overline{q_1}$. Hence show $\overline{q_2}\,\overline{q_1} = \overline{q_1 q_2}$.

8.7   A prime divisor property

It was shown in Section 8.5 that $\mathbb{Z}[\frac{1+i+j+k}{2}, i, j, k]$ has the division property, so this enables us to find the gcd of any two Hurwitz integers by the Euclidean algorithm.

However, since the quaternion product is generally noncommutative, we must distinguish between *right* and *left* divisors and stick to one type. We call δ a *right divisor* of α if $\alpha = \gamma\delta$ for some γ.

So if α and β have a common right divisor δ then

$$\alpha = \gamma\delta, \quad \beta = \varepsilon\delta \quad \text{for some } \gamma, \varepsilon,$$

and therefore

$$\rho = \alpha - \mu\beta = \gamma\delta - \mu\varepsilon\delta = (\gamma - \mu\varepsilon)\delta.$$

This shows that δ is also a right divisor of the remainder ρ when α is (right) divided by β.

Thus if we always divide on the right in the Euclidean algorithm, we obtain the greatest common *right* divisor of α and β. We call it the *right* gcd(α, β).

It then follows, by the usual inspection of terms produced by the Euclidean algorithm, that

$$\text{right gcd}(\alpha, \beta) = \mu\alpha + \nu\beta$$

for some Hurwitz integers μ and ν.

This allows us to prove the following prime divisor property (not the full analogue of those for $\mathbb{Z}$ and $\mathbb{Z}[i]$ but strong enough for our purposes): *if p is a real prime and if p divides a Hurwitz integer product $\alpha\beta$, then p divides α or p divides β.*

(It helps for p to be real because reals commute with all quaternions, hence p is both a right and left divisor of everything it divides.)

As usual, the proof begins by assuming that p does *not* divide α. Then

$$1 = \text{right gcd}(p, \alpha) = \mu p + \nu\alpha.$$

Multiplying both sides on the right by β gives

$$\beta = \mu p \beta + \nu\alpha\beta.$$

Since p divides $\mu p \beta$ (obviously) and $\nu\alpha\beta$ (by assumption), p divides the whole right-hand side. Hence p divides β, as required.   □

8.8   Proof of the four square theorem

We saw in Section 8.3 that the key to Lagrange's four square theorem is proving that every *prime* is the sum of four integer squares, since the four square identity takes care of all products of primes, that is, all other natural numbers except $1 = 0^2 + 0^2 + 0^2 + 1^2$.

The even prime $2 = 0^2 + 0^2 + 1^2 + 1^2$, so it remains to prove that any *odd* prime p is the sum of four integer squares. We do this with the help of the following proposition: *if $p = 2n + 1$, then there are $l, m \in \mathbb{Z}$ such that p divides $1 + l^2 + m^2$.*

This is analogous to Lagrange's lemma in Section 6.5, but easier. Here is the proof.

The squares x^2, y^2 of any two of the numbers $l = 0, 1, 2, \ldots, n$ are incongruent mod p because

$$x^2 \equiv y^2 \pmod{p} \Rightarrow x^2 - y^2 \equiv 0 \pmod{p}$$
$$\Rightarrow (x - y)(x + y) \equiv 0 \pmod{p}$$
$$\Rightarrow x \equiv y \text{ or } x + y \equiv 0 \pmod{p},$$

and $x + y \not\equiv 0 \pmod{p}$ since $0 < x + y < p$. Thus the $n + 1$ numbers $l = 0, 1, 2 \ldots, n$ give $n + 1$ incongruent values of l^2, mod p.

Similarly, the numbers $m = 0, 1, 2 \ldots, n$ give $n + 1$ incongruent values of m^2, hence of $-m^2$, and hence of $-1 - m^2$.

But only $2n+1$ incongruent values *exist*, mod $p = 2n+1$. Therefore, for some l and m we have

$$l^2 \equiv -1 - m^2 \quad (\bmod\ p).$$

That is, p divides $1 + l^2 + m^2$.   □

Four square theorem. *Every natural number is the sum of four squares.*

Proof. By the remarks above, it remains to prove the theorem for any odd prime p, which we have just shown to divide a number $1 + l^2 + m^2$.

To complete the proof we factorize $1 + l^2 + m^2$ into the product of Hurwitz integers

$$(1 - l\mathbf{i} - m\mathbf{j})(1 + l\mathbf{i} + m\mathbf{j})$$

and apply the prime divisor property from last section. If p is a Hurwitz prime, then p divides $1 - l\mathbf{i} - m\mathbf{j}$ or p divides $1 + l\mathbf{i} + m\mathbf{j}$. But neither conclusion is true, because neither

$$\frac{1}{p} - \frac{l\mathbf{i}}{p} - \frac{m\mathbf{j}}{p} \quad \text{nor} \quad \frac{1}{p} + \frac{l\mathbf{i}}{p} + \frac{m\mathbf{j}}{p}$$

is a Hurwitz integer. Hence *our arbitrary odd prime p is not a Hurwitz prime*, and therefore, by Section 8.6,

$$p = A^2 + B^2 + C^2 + D^2 \quad \text{with} \quad A, B, C, D, \in \mathbb{Z},$$

as required for the four square theorem.   □

Exercises

It follows from the four square theorem that any natural number has a Hurwitz integer factorization.

8.8.1 Explain why. (Does it matter if some of the squares are zero?)

Thus it is no longer any surprise that some real Gaussian primes are not Hurwitz primes—none of them are. However, we can still ask about the proper complex Gaussian primes $a + bi$ with $a, b \neq 0$.

8.8.2 Explain why the quaternions of the form $a + b\mathbf{i}$, for $a, b \in \mathbb{R}$, can be identified with the complex numbers $a + b\sqrt{-1}$.

8.8.3 Show that a proper Gaussian prime $a + bi$ is also a Hurwitz prime.

So far, we have said nothing about sums of three squares because their story is not so complete or elegant. For a start, there is no three square identity because a sum of three squares times a sum of three squares is not always a sum of three squares.

8.8.4 Find the natural numbers less than 20 that are not sums of three squares, and hence find one that factorizes into two sums of three squares.

8.8.5 Work out the possible values of x^2 mod 8, and hence show that no natural number of the form $8n+7$ is a sum of three squares.

With a little more work we can prove the more general result that no natural number of the form $4^m(8n+7)$ is a sum of three squares.

8.8.6 By considering the values of squares mod 4 show that

$$x^2+y^2+z^2 \equiv 0 \pmod 4$$

is possible only when x, y, z are all even.

8.8.7 Deduce from Exercise 8.8.6 that if $4^m(8n+7)$ is a sum of three squares, then so is $4^{m-1}(8n+7)$.

8.8.8 Exercises 8.8.7 and 8.8.5 imply that no natural number $4^m(8n+7)$ is a sum of three squares. Why?

The happy ending to this story is that the numbers $4^m(8n+7)$ are *precisely* those that are not sums of three squares. This was first proved by Legendre, and a proof may be found in Mordell (1969), pp. 175–178. As Mordell remarks "no really elementary treatment is known".

8.9   Discussion

Hurwitz' application of quaternions to the four square theorem was a historically natural event, very much like Dedekind's application of Gaussian integers to the two square theorem. In both cases a sum-of-squares identity was discovered first, followed considerably later by the discovery of generalized numbers with a multiplicative norm (the multiplicative property being just a restatement of the sum-of-squares identity). Finally, appropriate "integers" and "primes" among the generalized numbers are found to explain the representation of ordinary integers as sums of squares.

The historical parallel between the complex numbers $\mathbb{C}$ and the quaternions $\mathbb{H}$ is even stronger than this, because both stories have a similar missing link I have not yet mentioned. The discovery of the sum of squares identity led to the creation of the generalized numbers via an algebraic analysis of *rotations*.

In the case of complex numbers the story is briefly this.

- Diophantus (around 200 CE) observed the identity

$$(a_1^2 + b_1^2)(a_2^2 + b_2^2) = (a_1 a_2 - b_1 b_2)^2 + (a_1 b_2 + b_1 a_2)^2$$

  and interpreted it as a rule for taking two right-angled triangles, with side pairs (a_1, b_1) and (a_2, b_2), and generating a third triangle, with side pair $(a_1 a_2 - b_1 b_2, a_1 b_2 + b_1 a_2)$, whose hypotenuse is the product of those in the triangles (a_1, b_1) and (a_2, b_2).

- Viète (1593) noticed that the *angle* in the third triangle is the *sum* of the angles in the first two. In our notation, this is because the ratio of sides in the third triangle is

$$\frac{a_1 b_2 + b_1 a_2}{a_1 a_2 - b_1 b_2} = \frac{\frac{b_1}{a_1} + \frac{b_2}{a_2}}{1 - \frac{b_1}{a_1}\frac{b_2}{a_2}} = \tan(\theta_1 + \theta_2),$$

  where $\theta_1 = \tan^{-1}\frac{b_1}{a_1}$ and $\theta_2 = \tan^{-1}\frac{b_2}{a_2}$ are the angles in the first two.

- In the 18th century Cotes, de Moivre, and others rediscovered the angle-addition property by formally multiplying $\cos\theta + i\sin\theta$ and $\cos\phi + i\sin\phi$ to obtain $\cos(\theta + \phi) + i\sin(\theta + \phi)$. Multiplication of complex numbers of norm 1 therefore represents rotation of the plane about the origin. This, and the more obvious interpretation of addition as vector addition, led to the identification of complex numbers with points of the plane by Wessel (1797), Argand (1806), and (with more authority) by Gauss.

- Hamilton (1835) *defined* complex numbers to be pairs (a, b) of real numbers with addition and multiplication defined by

$$(a_1, b_1) + (a_2, b_2) = (a_1 + a_2, b_1 + b_2)$$
$$(a_1, b_1) \times (a_2, b_2) = (a_1 a_2 - b_1 b_2, a_1 b_2 + b_1 a_2).$$

Of course, Hamilton in 1835 was operating with 20/20 hindsight about the complex numbers, so he knew that these definitions of addition and multiplication would have all the usual algebraic properties, and that the function $\mathrm{norm}((a,b)) = a^2 + b^2$ would be multiplicative. However, he hoped that by rewriting the history of complex numbers in this way he would see *how to multiply triples*. He hoped in fact to find a multiplication rule for n-tuples that made their norm multiplicative, where

$$\mathrm{norm}((a_1, a_2, \ldots, a_n)) = a_1^2 + a_2^2 + \cdots + a_n^2.$$

But a multiplicative norm for n-tuples implies a sum of n squares identity, so it would have been wise to look for a sum of three squares identity first.

This did not happen. Instead, Hamilton spent 13 years trying in vain to find a multiplication rule for triples. Virtually all that he learned from his search was that the commutative law of multiplication might have to be abandoned. When he also abandoned triples, and tried quadruples, everything fell into place. On October 16 1843 he wrote down the rules

$$\mathbf{i}^2 = \mathbf{j}^2 = \mathbf{k}^2 = \mathbf{ijk} = -1$$

that define quaternion multiplication and from them derived the four square identity. Only then did he start to catch up on the news—that Euler knew the four square identity in 1748, that Legendre knew that there is no three square identity, and that quaternion multiplication had already been used by Rodrigues in 1840 to compute the product of rotations in $\mathbb{R}^3$.

Of course, these earlier discoveries were mere glimpses of the complete and beautiful structure discovered by Hamilton. The quaternions are even more remarkable than he knew, because after his death it was shown that "multiplying n-tuples" is possible only for $n = 1, 2, 4$, and 8. To be precise, these are the only n for which $\mathbb{R}^n$ has a multiplication that distributes over vector addition, and a multiplicative norm. A related result, due to Hurwitz, is that an n square identity exists only for $n = 1, 2, 4$, and 8.

For $n = 1, 2, 4$ the corresponding structures are $\mathbb{R}$, $\mathbb{C}$, $\mathbb{H}$, and for $n = 8$ the corresponding structure is called the *octonions*. It was discovered by Hamilton's friend John Graves, just months after the discovery of quaternions, and is based on an eight square identity. Like the quaternions, the octonions do not have commutative multiplication; their multiplication is not associative either. More on these generalized number systems may be found in the excellent book *Numbers* by Ebbinghaus *et al.* (1991).

Like the quaternions themselves, the Hurwitz integers had an interesting precursor in geometry. In 1852 Schläfli discovered that there are two exceptional dimensions n for which $\mathbb{R}^n$ can be "tiled" by regular figures other than cubes. They are $n = 2$, where the exceptional tilings are by equilateral triangles or by regular hexagons, and $n = 4$. In $\mathbb{R}^2 = \mathbb{C}$ the two exceptional tilings can both be derived from the Eisenstein integers $\mathbb{Z}[\zeta_3]$. The triangle tiling is obtained by joining each integer point to its nearest neighbors and the hexagon tiling by taking each integer point as the center of a region whose sides are midway between neighboring integer points. In $\mathbb{R}^4$ the two exceptional tilings are obtained in the same way, from none other than the Hurwitz integers. For more on these remarkable tilings, see Coxeter (1948).

9

Quadratic reciprocity

PREVIEW

Fermat's remarkable discovery that odd primes of the *quadratic* form $x^2 + y^2$ are in fact those of the *linear* form $4n + 1$ led to the more general problem of describing primes of the form $x^2 + dy^2$ for nonsquare integers d. Is it true, for each d, that the primes of the form $x^2 + dy^2$ are those of a finite number of linear forms?

Fermat found such forms for the primes $x^2 + 2y^2$ and $x^2 + 3y^2$ as well. In each case a crucial step in determining the linear forms of the primes $x^2 + dy^2$ is to find the *quadratic character* of $-d$, that is, to find the primes q such that $-d$ is a square, mod q.

The *law of quadratic reciprocity* answers all such questions. The law describes when p is a square, mod q, for odd primes p and q; and its *supplements* deal with the cases $p = -1$ and $p = 2$.

To prove it, we first prove *Euler's criterion*, which states that p is a square mod $q \Leftrightarrow p^{\frac{q-1}{2}} \equiv 1 \pmod{q}$. This yields the supplements fairly easily, and it also helps in the proof of the law itself.

We also need the *Chinese remainder theorem*. It is of interest in itself and for what it says about the Euler φ function, but our main purpose is to use it to prove quadratic reciprocity for odd primes p and q.

To discuss quadratic reciprocity tersely we use *Legendre's symbol* $\left(\frac{p}{q}\right)$, which equals 1 when P is a square mod q, and -1 otherwise. All values of $\left(\frac{p}{q}\right)$ follow from the values $\left(\frac{p}{q}\right)$ for odd primes p (from quadratic reciprocity) and the special values $\left(\frac{-1}{q}\right)$ and $\left(\frac{2}{q}\right)$ (from the supplements).

158

9.1   Primes $x^2 + y^2$, $x^2 + 2y^2$, and $x^2 + 3y^2$

The primes $x^2 + y^2$ again

In the proof of Fermat's two square theorem, that an odd prime p is of the form $x^2 + y^2$ if and only if p is of the form $4n + 1$, a key step was showing that any prime $p = 4n + 1$ divides a number of the form $m^2 + 1$. We proved this in Section 6.5 using Wilson's theorem to construct a suitable m.

We now re-examine this step to see how it might be generalized. The statement that p divides $m^2 + 1$ is equivalent to

$$-1 \equiv m^2 \quad (\text{mod } p),$$

in other words -1 *is a square*, mod $p = 4n + 1$. And indeed our proof was to take the expression for -1 given by Wilson's theorem and show that it was in fact a square, mod $p = 4n + 1$.

This raises the general question of whether q is a square mod p, where p and q are arbitrary integers. We also state the question as: what is the *quadratic character* of q, mod p? Several problems lead to this question, as we now show.

The form $x^2 + 2y^2$

After describing the primes of the form $x^2 + y^2$, Fermat tackled primes of the form $x^2 + 2y^2$. He claimed that

$$p = x^2 + 2y^2 \Leftrightarrow p = 8n + 1 \text{ or } p = 8n + 3.$$

A proof can be given along the same lines as our proof of the two square theorem.

We work in $\mathbb{Z}[\sqrt{-2}]$, and prove first that *if p is an ordinary prime that is not a prime of $\mathbb{Z}[\sqrt{-2}]$ then*

$$p = a^2 + 2b^2 \quad \text{for some } a, b \in \mathbb{Z}.$$

The proof is like that for non-Gaussian primes (Section 6.3). If p is not a prime of $\mathbb{Z}[\sqrt{-2}]$ then it has factors of norm > 1, say

$$p = (a + b\sqrt{-2})\gamma.$$

Multiplying this equation by its conjugate leads to $p = a^2 + 2b^2$.

The key step now is to prove that *any prime $p = 8n+1$ or $8n+3$ divides a number of the form $m^2 + 2$*. Once this is done, one uses the factorization

$$m^2 + 2 = (m - \sqrt{-2})(m + \sqrt{-2})$$

and unique prime factorization in $\mathbb{Z}[\sqrt{-2}]$ to complete the proof as we did for $\mathbb{Z}[i]$ (Section 6.5).

The claim that p divides a number of the form $m^2 + 2$ is equivalent to

$$-2 \equiv m^2 \pmod{p}.$$

Thus we have to prove that -2 *is a square*, mod p, when p is a prime of the form $8n + 1$ or $8n + 3$.

The form $x^2 + 3y^2$

Next, Fermat described the primes of the form $x^2 + 3y^2$:

$$p = x^2 + 3y^2 \Leftrightarrow p = 3n + 1.$$

This can be proved along the same lines as for $x^2 + y^2$ and $x^2 + 2y^2$, this time using factorizations in $\mathbb{Z}[\frac{-1+\sqrt{-3}}{2}]$. The awkward step is to prove that any prime $p = 3n + 1$ divides a number of the form $m^2 + 3$. Equivalently,

$$-3 \equiv m^2 \pmod{p},$$

so we now have to prove that -3 *is a square*, mod $p = 3n + 1$.

Exercises

In a letter to Frenicle on 15 June 1641, Fermat asked which natural numbers are the sums of the two smaller members of a Pythagorean triple. These are the numbers of the form $2XY + X^2 - Y^2 = (X+Y)^2 - 2Y^2$, and Frenicle correctly replied that the primes among the numbers $x^2 - 2y^2$ are precisely those of the form $8n \pm 1$. This can be proved in the same way as Fermat's results about $x^2 + y^2$, $x^2 + 2y^2$, $x^2 + 3y^2$ by using

- conjugation in $\mathbb{Z}[\sqrt{2}]$,

- the quadratic character of 2,

- unique prime factorization in $\mathbb{Z}[\sqrt{2}]$.

The quadratic character of 2 will be established in Section 9.4, but the remaining steps can be done here.

9.1.1 Suppose p is an ordinary prime but not prime in $\mathbb{Z}[\sqrt{2}]$, so $p = (a+b\sqrt{2})\gamma$, where $a+b\sqrt{2}$ and γ have norm of absolute value greater than 1. By taking conjugates of both sides, show that $p = a^2 - 2b^2$.

9.1.2 Using the fact that $x^2 \equiv 0, 1, 4 \pmod{8}$ show that all odd primes of the form $x^2 - 2y^2$ are of the form $8n \pm 1$.

One now uses the quadratic character of 2 (see Section 9.4) to prove that any prime of the form $8n \pm 1$ divides a natural number of the form $m^2 - 2$. Assuming also a prime divisor property in $\mathbb{Z}[\sqrt{2}]$, the argument continues as follows.

9.1.3 Show that, if p divides $m^2 - 2 = (m - \sqrt{2})(m + \sqrt{2})$, then p is not a prime of $\mathbb{Z}[\sqrt{2}]$. (Hence p is of the form $x^2 - 2y^2$ by Exercise 9.1.1.)

The only information now missing, apart from the quadratic character of 2, is a proof that $\mathbb{Z}[\sqrt{2}]$ has the prime divisor property. This is obtained by showing that $\mathbb{Z}[\sqrt{2}]$ has the division property, and hence a Euclidean algorithm.

The norm $a^2 - 2b^2$ of $a + b\sqrt{2} \in \mathbb{Z}[\sqrt{2}]$ is *not* $|a+b\sqrt{2}|^2$, so the geometric argument used for $\mathbb{Z}[i]$ and $\mathbb{Z}[\sqrt{-2}]$ does not apply, and we opt for a purely algebraic approach. First we state the division property of $\mathbb{Z}[\sqrt{2}]$ as follows.

If $\alpha, \beta \in \mathbb{Z}[\sqrt{2}]$ and $\beta \neq 0$, then there are $\mu, \rho \in \mathbb{Z}[\sqrt{2}]$ with

$$\alpha = \mu\beta + \rho \quad and \quad |\mathrm{norm}(\rho)| < |\mathrm{norm}(\beta)|.$$

9.1.4 Show that the division property is implied by the existence of a $\mu \in \mathbb{Z}[\sqrt{2}]$ with $\left|\mathrm{norm}\left(\frac{\alpha}{\beta} - \mu\right)\right| < 1$. (We are now extending the norm to $\mathbb{Q}(\sqrt{2})$. Is this OK?)

9.1.5 If $\alpha, \beta \in \mathbb{Z}[\sqrt{2}]$ and $\beta \neq 0$, show by "rationalizing the denominator" that

$$\frac{\alpha}{\beta} = \frac{A_1}{\mathrm{norm}(\beta)} + \frac{A_2}{\mathrm{norm}(\beta)}\sqrt{2} \quad \text{for some } A_1, A_2 \in \mathbb{Z}.$$

9.1.6 Continuing with the notation of Exercise 9.1.5, if

$$m_1 = \text{nearest integer to } \frac{A_1}{\mathrm{norm}(\beta)}, \quad m_2 = \text{nearest integer to } \frac{A_2}{\mathrm{norm}(\beta)},$$

and $\mu = m_1 + m_2\sqrt{2}$, show that $\left|\mathrm{norm}\left(\frac{\alpha}{\beta} - \mu\right)\right| < 1$, so $\mathbb{Z}[\sqrt{2}]$ has the division property.

9.2 Statement of quadratic reciprocity

In the mid-18th century Euler realized that knowing the primes of forms such as $x^2 + y^2$, $x^2 + 2y^2$ and $x^2 + 3y^2$ depends on knowing whether p is a square mod q, for certain integers p and q. In the case where p and q are both odd primes he conjectured that the answer is:

When p and q are both of the form $4n + 3$, then

$$p \text{ is a square, mod } q \Leftrightarrow q \text{ is } not \text{ a square, mod } p.$$

Otherwise

$$p \text{ is a square, mod } q \Leftrightarrow q \text{ is a square, mod } p.$$

Because of the reciprocal relationship between p and q, this statement is called the law of *quadratic reciprocity*. (The word "quadratic" in this case really means "square". In the literature one often finds the old term "quadratic residues mod p" for "squares mod p".)

Euler was unable to prove the law of quadratic reciprocity. The first proofs were given by Gauss in 1801. Since then nearly 200 different proofs have been given, making quadratic reciprocity the second most proved theorem in mathematics, after Pythagoras' theorem.

Notation and examples

In Section 9.8 we give a recent proof of quadratic reciprocity, which simplifies one of Gauss. But first we introduce some notation and give examples of its use.

For any primes p and q the *Legendre symbol* or *quadratic character symbol* is defined by

$$\left(\frac{p}{q} \right) = \begin{cases} 1 & \text{if } p \text{ is a square mod } q \\ -1 & \text{if } p \text{ is not a square mod } q \end{cases}$$

With the help of this symbol, quadratic reciprocity can be stated very concisely as

$$\left(\frac{p}{q} \right) \left(\frac{q}{p} \right) = (-1)^{\frac{p-1}{2} \frac{q-1}{2}}$$

for odd primes p and q.

The Legendre symbol may be extended to $\left(\frac{P}{q} \right)$ for any integer P, and it is ± 1 according as P is a square mod q or not. This is possible by the *multiplicative property*

$$\left(\frac{P}{q} \right) = \left(\frac{p_1 p_2 \cdots p_k}{q} \right) = \left(\frac{p_1}{q} \right) \left(\frac{p_2}{q} \right) \cdots \left(\frac{p_k}{q} \right)$$

where $p_1 p_2 \cdots p_k$ is the prime factorization of P (possibly including the prime 2 and the unit -1). To evaluate the possible factors $\left(\frac{-1}{q}\right)$ and $\left(\frac{2}{q}\right)$ arising from this prime factorization, we need the so-called *supplements to quadratic reciprocity*,

$$\left(\frac{-1}{q}\right) = 1 \Leftrightarrow q = 4n+1, \qquad\qquad \text{(I)}$$

$$\left(\frac{2}{q}\right) = 1 \Leftrightarrow q = 8n \pm 1. \qquad\qquad \text{(II)}$$

These, and the multiplicative property, are proved in the next few sections. We now use them to prove properties of squares sought in Section 9.1.

Examples

To show that -2 is a square mod $p = 8n+1$ or $8n+3$ we calculate

$$\left(\frac{-2}{8n+1}\right) = \left(\frac{-1}{8n+1}\right)\left(\frac{2}{8n+1}\right) \quad \text{by the multiplicative property}$$

$$= 1 \times 1 = 1 \quad \text{by supplements (I) and (II)}.$$

$$\left(\frac{-2}{8n+3}\right) = \left(\frac{-1}{8n+3}\right)\left(\frac{2}{8n+3}\right) \quad \text{by the multiplicative property}$$

$$= (-1) \times (-1) = 1 \quad \text{by supplements (I) and (II)}.$$

To show that -3 is a square mod $p = 3n+1$:

$$\left(\frac{-3}{3n+1}\right) = \left(\frac{-1}{3n+1}\right)\left(\frac{3}{3n+1}\right) \quad \text{by the multiplicative property}$$

$$= 1 \times \left(\frac{3n+1}{3}\right) \text{ or } (-1) \times (-1)\left(\frac{3n+1}{3}\right),$$

$$\text{by supplement (I) and quadratic reciprocity,}$$

with the $+$ or $-$ sign according as $3n+1$ is of the form $4n'+1$ or not,

$$= 1 \times \left(\frac{1}{3}\right) \text{ or } (-1) \times (-1)\left(\frac{1}{3}\right) \quad \text{since } 3n+1 \equiv 1 \pmod 3$$

$$= 1 \times 1 \text{ or } (-1) \times (-1) = 1, \quad \text{since 1 is a square mod 3.}$$

Exercises

Note that the quadratic character of 2, given by supplement (II), is precisely what we need to fill the gap in the last exercise set, proving that the odd primes of the form $x^2 - 2y^2$ are those of the form $8n \pm 1$. We now use the quadratic character of 3 in a similar way to characterize the primes of the form $x^2 - 3y^2$.

9.2.1 Suppose p is an ordinary prime but not prime in $\mathbb{Z}[\sqrt{3}]$, so $p = (a + b\sqrt{3})\gamma$, where $a + b\sqrt{3}$ and γ have norm of absolute value greater than 1. By taking conjugates of both sides, show that $p = a^2 - 3b^2$.

9.2.2 Use congruence mod 12 to show that all odd primes of the form $x^2 - 3y^2$ are of the form $12n + 1$.

9.2.3 Use quadratic reciprocity to show that 3 is a square mod any prime $p = 12n + 1$, and hence that such a p divides a natural number of the form $m^2 - 3$.

9.2.4 Check that the argument of Exercise 9.1.6 also works for $\mathbb{Z}[\sqrt{3}]$, so $\mathbb{Z}[\sqrt{3}]$ has the prime divisor property.

9.2.5 Use Exercises 9.2.3 and 9.2.4, and $m^2 - 3 = (m - \sqrt{3})(m + \sqrt{3})$, to prove that $p = 12n + 1$ is not a prime of $\mathbb{Z}[\sqrt{3}]$. Conclude, by Exercise 9.2.1, that p is of the form $x^2 - 3y^2$.

Thus the form $x^2 - 3y^2$ represents all the odd primes of the form $12n + 1$. It does not represent the even prime 2, as this would contradict the quadratic character of 2.

9.2.6 Show that an integer solution of $x^2 - 3y^2 = 2$ implies that 2 is a square modulo a prime not allowed by supplement (II).

Likewise, the quadratic character of -1 gives another solution of Exercise 5.8.6.

9.2.7 Show that an integer solution of $x^2 - 3y^2 = -1$ implies that -1 is a square modulo a prime not allowed by supplement (I).

9.3   Euler's criterion

If q is prime and $a \not\equiv 0 \pmod{q}$, then $a^{q-1} \equiv 1 \pmod{q}$ by Fermat's little theorem. Euler used this to derive the following formula:

Euler's criterion. *For an odd prime q, $\left(\dfrac{p}{q}\right) \equiv p^{\frac{q-1}{2}} \pmod{q}$, and hence p is a square mod $q \Leftrightarrow p^{\frac{q-1}{2}} \equiv 1 \pmod{q}$.*

Proof. First suppose that p is a square mod q, say $p \equiv a^2 \pmod{q}$. Then $\left(\frac{p}{q}\right) = 1$ by definition and

$$p^{\frac{q-1}{2}} \equiv a^{q-1} \equiv 1 \pmod{q} \quad \text{by Fermat's little theorem.}$$

Conversely, if p is not a square mod q, it suffices to show that

$$p^{\frac{q-1}{2}} \not\equiv 1 \pmod{q}.$$

This is so because $x = p^{\frac{q-1}{2}}$ satisfies $x^2 \equiv p^{q-1} \equiv 1 \pmod{q}$ by Fermat's little theorem, and $x^2 \equiv 1$ has only the two solutions $x \equiv \pm 1$ by Lagrange's polynomial congruence theorem.

By the same theorem, $p^{\frac{q-1}{2}} \equiv 1 \pmod{q}$ has at most $\frac{q-1}{2}$ solutions, and we know that they include the squares $p = 1^2, 2^2, \ldots, \left(\frac{q-1}{2}\right)^2$. These $\frac{q-1}{2}$ squares are distinct. Indeed, if x^2 and y^2 are any two of them we have

$$x^2 \equiv y^2 \pmod{q} \Rightarrow x^2 - y^2 \equiv 0 \pmod{q}$$
$$\Rightarrow (x-y)(x+y) \equiv 0 \pmod{q}$$
$$\Rightarrow x = y.$$

This is so because $1 < x+y < q$ and hence $x+y \not\equiv 0 \pmod{q}$.

Thus when $p \not\equiv a^2 \pmod{q}$ we have $p^{\frac{q-1}{2}} \not\equiv 1 \pmod{q}$ and therefore $p^{\frac{q-1}{2}} \equiv -1 \equiv \left(\frac{p}{q}\right) \pmod{q}$. □

Notice that the proof of this criterion does not assume that p is actually prime. We take this opportunity to define $\left(\frac{P}{q}\right)$, for any $P \not\equiv 0 \pmod{q}$, to be 1 if P is a square mod q and -1 otherwise. Then the Euler criterion gives an easy proof of the following.

Multiplicative property of $\left(\frac{P}{q}\right)$. *For any $P_1, P_2 \not\equiv 0 \pmod{q}$*

$$\left(\frac{P_1}{q}\right)\left(\frac{P_2}{q}\right) = \left(\frac{P_1 P_2}{q}\right).$$

Proof. By Euler's criterion

$$\left(\frac{P_1}{q}\right) \equiv P_1^{\frac{q-1}{2}} \pmod{q},$$

$$\left(\frac{P_2}{q}\right) \equiv P_2^{\frac{q-1}{2}} \pmod{q},$$

and therefore

$$\left(\frac{P_1}{q}\right)\left(\frac{P_2}{q}\right) \equiv P_1^{\frac{q-1}{2}} P_2^{\frac{q-1}{2}} \pmod{q} \quad \text{by multiplication of congruences}$$

$$\equiv (P_1 P_2)^{\frac{q-1}{2}} \pmod{q}$$

$$\equiv \left(\frac{P_1 P_2}{q}\right) \pmod{q} \quad \text{by Euler's criterion again.} \qquad \square$$

The proof of the multiplicative property also does not assume that the integers P are prime. So we can evaluate $\left(\frac{P}{q}\right)$ for any integer P, provided we know $\left(\frac{p}{q}\right)$ for factors p of P. We can assume that the factors are among $-1, 2$ and the odd primes.

The law of quadratic reciprocity (proved in Section 9.8) gives information about $\left(\frac{p}{q}\right)$ for odd primes p, so we also need information about $\left(\frac{-1}{q}\right)$ and $\left(\frac{2}{q}\right)$. We obtain this from the *supplements to quadratic reciprocity*, proved here and in the next section, which give the values of $\left(\frac{-1}{q}\right)$ and $\left(\frac{2}{q}\right)$ directly. (We previously gave another determination of $\left(\frac{-1}{q}\right)$ in Section 6.7.)

The value of $\left(\frac{-1}{q}\right)$. *For an odd prime q*

$$\left(\frac{-1}{q}\right) = \begin{cases} 1 \text{ if } q = 4n+1 \\ -1 \text{ if } q = 4n+3. \end{cases}$$

Proof. Euler's criterion says that

$$\left(\frac{-1}{q}\right) \equiv (-1)^{\frac{q-1}{2}} \pmod{q}.$$

Thus if $q = 4n+1$ we have

$$\left(\frac{-1}{q}\right) \equiv (-1)^{2n} \equiv 1 \pmod{q},$$

and if $q = 4n+3$ we have

$$\left(\frac{-1}{q}\right) \equiv (-1)^{2n+1} \equiv -1 \pmod{q}. \qquad \square$$

Exercises

If one assumes the existence of a primitive root for a prime q (proved in Section 3.9), then it is possible to give a simpler proof of Euler's criterion.

9.3.1 If a is a primitive root for q, so that $1, a, a^2, \ldots, a^{q-2}$ are the distinct nonzero elements mod q, show that the squares mod q are $1, a^2, a^4, \ldots, a^{q-3}$.

9.3.2 Deduce from Exercise 9.3.1 that b is a square, mod $q \Leftrightarrow b^{\frac{q-1}{2}} \equiv 1 \pmod{q}$.

Another easy consequence of Exercise 9.3.1 is a "half and half" property of squares mod q.

9.3.3 Show that exactly half of $1, 2, 3, \ldots, q-1$ are squares mod q.

The "half and half" property can also be proved without assuming the existence of a primitive root, though not quite so easily. The proof of Euler's criterion shows that at least half of $1, 2, \ldots, q-1$ are squares. Then it suffices to prove the following:

9.3.4 Show that $1^2, 2^2, 3^2, \ldots, (q-1)^2$ include at most half of the numbers $\neq 0$ (mod q).

9.4 The value of $\left(\dfrac{2}{q}\right)$

Euler's criterion says that $\left(\frac{2}{q}\right) \equiv 2^{\frac{q-1}{2}} \pmod{q}$, but $2^{\frac{q-1}{2}}$ is harder to evaluate than $(-1)^{\frac{q-1}{2}}$. Fermat seems to have known $\left(\frac{2}{q}\right)$ (see exercises to Section 9.1), but we do not know how. It turns out that

$$2^{\frac{q-1}{2}} \equiv \begin{cases} (-1)^{\frac{q-1}{4}} \pmod{q} & \text{if } q = 4n+1 \\ (-1)^{\frac{q+1}{4}} \pmod{q} & \text{if } q = 4n+3. \end{cases}$$

We can prove this by manipulating the product $1 \times 2 \times 3 \times \cdots \times (q-1)$ (mod q), a little like the manipulation in Section 6.5 that yielded the quadratic character of -1 in Section 6.7.

When $q = 4n+1$ the manipulation takes 2 out of half of the factors, -1 out of one quarter of them, and rewrites the resulting negative factors mod $4n+1$ to make them positive. This restores the product we started with, which can then be cancelled from both sides.

$$1 \times 2 \times \cdots \times 4n$$

$$\equiv (1 \times 3 \times \cdots \times (4n-1)) \times (2 \times 4 \times \cdots \times 4n) \qquad (\bmod\ q)$$

$$\equiv (1 \times 3 \times \cdots \times (4n-1)) \times (1 \times 2 \times \cdots \times 2n)2^{2n} \qquad (\bmod\ q)$$

$$\equiv (1 \times 3 \times \cdots \times (2n-1)) \times ((2n+1)(2n+3)\cdots(4n-1))$$
$$\times (1 \times 2 \times \cdots \times 2n)2^{2n} \qquad (\bmod\ q)$$

$$\equiv ((-1)(-3)\cdots(-2n+1))(-1)^n \times ((2n+1)(2n+3)\cdots(4n-1))$$
$$\times (1 \times 2 \times \cdots \times 2n)2^{2n} \qquad (\bmod\ q)$$

$$\equiv ((4n)(4n-2)\cdots(2n+2))(-1)^n \times ((2n+1)(2n+3)\cdots(4n-1))$$
$$\times (1 \times 2 \times \cdots \times 2n)2^{2n} \qquad (\bmod\ q)$$

$$\text{since } -1 \equiv 4n,\ -3 \equiv 4n-2,\ \ldots \qquad (\bmod\ q)$$

$$\equiv ((2n+1)(2n+2)\cdots(4n))(-1)^n \times (1 \times 2 \times \cdots \times 2n)2^{2n} \quad (\bmod\ q)$$

$$\equiv (-1)^n 2^{2n}(1 \times 2 \times \cdots \times 4n) \qquad (\bmod\ q)$$

Cancelling $1 \times 2 \times \cdots \times 4n$ from the first and last line, we get

$$1 \equiv (-1)^n 2^{2n} \equiv (-1)^{\frac{q-1}{4}} 2^{\frac{q-1}{2}} \quad (\bmod\ q),$$

that is,

$$2^{\frac{q-1}{2}} \equiv (-1)^{\frac{q-1}{4}} \quad (\bmod\ q), \quad \text{when } q = 4n+1.$$

There is a similar proof (exercises) that

$$2^{\frac{q-1}{2}} \equiv (-1)^{\frac{q+1}{4}} \quad (\bmod\ q), \quad \text{when } q = 4n+3.$$

To decide when 2 is a square mod q we therefore have to look at two cases:

If $q = 4n+1$, $\frac{q-1}{4} = n$. So $\left(\frac{2}{q}\right) = (-1)^{\frac{q-1}{4}}$ is 1 when $n = 2m$ and -1 when $n = 2m+1$. That is, 2 is a square mod q when $q = 8m+1$, not when $q = 8m+5$.

If $q = 4n+3$, $\frac{q+1}{4} = n+1$. So $\left(\frac{2}{q}\right) = (-1)^{\frac{q+1}{4}}$ is 1 when $n = 2m+1$ and -1 when $n = 2m$. That is, 2 is a square mod q when $q = 8m+7$, not when $q = 8m+3$.

To sum up: 2 *is a square* mod $q \Leftrightarrow q = 8m \pm 1$.

Exercises

The proof for $q = 4n+3$ splits the product $1 \times 2 \times \cdots \times 4n \times (4n+1) \times (4n+2)$ in a slightly less regular way—necessarily, because the number of terms is not divisible by 4. The way to do it can be seen by trying an example first, say $q = 11$.

9.4.1 By partitioning $10! = 1 \times 2 \times 3 \times 4 \times 5 \times 6 \times 7 \times 8 \times 9 \times 10$ as

$$(1 \times 3 \times 5)(7 \times 9)(2 \times 4 \times 6 \times 8 \times 10),$$

removing -1 and 2 from appropriate factors, and then changing negative factors $-k$ back to positive $11-k$, show that

$$10! \equiv 10!(-1)^3 2^5 \pmod{11},$$

and hence that $\left(\frac{2}{11}\right) = -1$.

9.4.2 By partitioning $(4n+2)! = 1 \times 2 \times 3 \times \cdots \times 4n \times (4n+1) \times (4n+2)$ as

$$(1 \times 3 \times 5 \times \cdots \times (2n+1))((2n+3) \times (2n+5) \times \cdots \times (4n+1))$$
$$\times (2 \times 4 \times 6 \times \cdots \times 4n \times (4n+2)),$$

removing -1 and 2 from appropriate factors, and then changing negative factors $-k$ back to positive $4n+3-k$, show that

$$(4n+2)! \equiv (4n+2)!(-1)^{n+1} 2^{2n+1} \pmod{4n+3},$$

and hence that $\left(\frac{2}{4n+3}\right) = (-1)^{n+1}$.

9.4.3 Deduce from Exercise 9.4.2 that $\left(\frac{2}{q}\right) = (-1)^{\frac{q+1}{4}}$ when $q = 4n+3$.

9.5   The story so far

In Section 9.1 we observed that classifying primes of the forms $x^2 + y^2$, $x^2 + 2y^2$, $x^2 + 3y^2$ depends on knowing that certain numbers are squares mod certain primes. To prove such results we introduced the *Legendre symbol*, defined for any integer $P \not\equiv 0 \pmod{q}$ and odd primes q by

$$\left(\frac{P}{q}\right) = \begin{cases} 1 & \text{if } P \text{ is a square, mod } q \\ -1 & \text{if } P \text{ is not a square, mod } q. \end{cases}$$

Thanks to the *Euler criterion*,

$$\left(\frac{P}{q}\right) \equiv P^{\frac{q-1}{2}} \pmod{q},$$

valid for any $P \not\equiv 0 \pmod{q}$, we can prove the *multiplicative property*

$$\left(\frac{P_1}{q}\right)\left(\frac{P_2}{q}\right) = \left(\frac{P_1 P_2}{q}\right),$$

and hence find $\left(\frac{P}{q}\right)$ for any $P \not\equiv 0 \pmod{q}$ by splitting P into factors $P_1, P_2, \ldots$ that are either -1 or primes, then multiplying $\left(\frac{P_1}{q}\right), \left(\frac{P_2}{q}\right), \ldots$.

In Sections 9.3 and 9.4 we used the Euler criterion to prove the *supplements to quadratic reciprocity*:

$$\left(\frac{-1}{q}\right) = 1 \Leftrightarrow q = 4n+1 \tag{I}$$

$$\left(\frac{2}{q}\right) = 1 \Leftrightarrow q = 8n \pm 1. \tag{II}$$

Thus it remains to evaluate $\left(\frac{p}{q}\right)$ for odd primes p and q. This is done by the *quadratic reciprocity law*, which is proved in Section 9.8:

$$\left(\frac{p}{q}\right)\left(\frac{q}{p}\right) = (-1)^{\frac{p-1}{2}\frac{q-1}{2}}.$$

Using quadratic reciprocity

Quadratic reciprocity says that

$$\left(\frac{p}{q}\right) = \left(\frac{q}{p}\right) \qquad \text{if one of } p, q = 4n+1,$$

$$\left(\frac{p}{q}\right) = -\left(\frac{q}{p}\right) \qquad \text{otherwise.}$$

Another point to bear in mind is that if $p \equiv p' \pmod{q}$, then

$$p \text{ is a square mod } q \Leftrightarrow p' \text{ is a square mod } q.$$

Thus we can replace p in $\left(\frac{p}{q}\right)$ by its remainder p' on division by q. One then "reciprocates" $\left(\frac{p'}{q}\right)$ to $\pm\left(\frac{q}{p'}\right)$ by quadratic reciprocity, replaces q by its remainder q' on division by p', and so on. In effect, one interweaves the Euclidean algorithm with applications of multiplicativity to rapidly reduce the numbers to the point where supplements I and II can be applied.

Example. Decide whether 37 is a square mod 59.

$$\left(\frac{37}{59}\right) = \left(\frac{59}{37}\right) \quad \text{by reciprocity}$$

$$= \left(\frac{22}{37}\right) \quad \text{by remaindering}$$

$$= \left(\frac{2}{37}\right)\left(\frac{11}{37}\right) \quad \text{by multiplicativity}$$

$$= -\left(\frac{11}{37}\right) \quad \text{by supplement (II)}$$

$$= -\left(\frac{37}{11}\right) \quad \text{by reciprocity}$$

$$= -\left(\frac{4}{11}\right) \quad \text{by remaindering}$$

$$= -\left(\frac{2}{11}\right)^2 \quad \text{by multiplicativity}$$

$$= -1$$

Hence 37 is *not* a square mod 59.

Exercises

9.5.1 Show that $\left(\frac{55}{89}\right) = 1$ by using multiplicativity and the Euclidean algorithm.

9.5.2 Verify directly that $\left(\frac{55}{89}\right) = 1$ by finding a square $\equiv 55 \pmod{89}$.

9.5.3 Show that $\left(\frac{56}{89}\right) = -1$.

9.6   The Chinese remainder theorem

An example

The Chinese remainder theorem is about representing numbers by their remainders. For example, here are the numbers $n = 0, 1, 2, \ldots, 14$ and their remainders $n \bmod 3$ and $n \bmod 5$ on division by 3 and 5 respectively.

n	0	1	2	3	4	5	6	7	8	9	10	11	12	13	14
$n \bmod 3$	0	1	2	0	1	2	0	1	2	0	1	2	0	1	2
$n \bmod 5$	0	1	2	3	4	0	1	2	3	4	0	1	2	3	4

It can be checked that each of the 15 numbers $n = 0, 1, 2 \ldots, 14$ has a different pair of remainders, and hence *each such n is determined by its pair of remainders*. For example, the only number with the pair of remainders $(2,3)$ is 8.

It is also easy to see why this is true.

- The first component of each pair, $n \bmod 3$, runs through the sequence $012012\ldots$, which repeats every three steps.

- The second component, $n \bmod 5$, runs through $0123401234\ldots$, which repeats every five steps.

- Therefore, no pair is repeated until after $\mathrm{lcm}(3,5) = 15$ steps, and hence the first 15 pairs are different.

Classical Chinese remainder theorem

The original form of the remainder theorem, found in China around 300 CE, goes as follows. *If* $\gcd(a,b) = 1$*, then each* $n = 0, 1, 2, \ldots, ab - 1$ *has a distinct pair of remainders on division by a and b.*

This can be proved by a generalization of the argument above.

- The first remainder of each pair, $n \bmod a$, runs through the sequence

$$012\ldots(a-1)012\ldots(a-1)\ldots$$

which repeats every a steps.

- The second remainder, $n \bmod b$, runs through the sequence

$$012\ldots(b-1)012\ldots(b-1)\ldots$$

which repeats every b steps.

- Therefore, no remainder pair is repeated until after $\mathrm{lcm}(a,b) = ab$ steps, and hence the first ab pairs are different.   □

The condition $\gcd(a,b) = 1$ says that a and b have no common prime factor, so their common multiples include all their prime factors, and hence $\mathrm{lcm}(a,b) = ab$.

Exercises

The classical example of a Chinese remainder problem is in the *Mathematical Manual* of Sun Zi, late in the 3rd century CE. It is required to find a number that leaves remainder 2 on division by 3, remainder 3 on division by 5, and remainder 2 on division by 7.

9.6.1 Show that the numbers $1, 2, 3, \ldots, 210$ all leave distinct triples of remainders on division by 3, 5, 7 respectively.

9.6.2 Find a generalization of this result to triples of remainders on division by a, b, c, with suitable conditions on the moduli a, b, c.

9.6.3 Find the minimal solution of Sun Zi's problem.

9.6.4 Describe the numbers with remainder 1 on division by 3 and remainder 2 on division by 5, and hence find the least of them with remainder 3 on division by 7.

9.7   The full Chinese remainder theorem

The modern form of the theorem not only represents each of the numbers $n = 0, 1, 2, \ldots ab - 1$ by a pair $(n \bmod a, n \bmod b)$, it also recognizes that these n can be *added and multiplied* by adding and multiplying the corresponding pairs.

The first components of pairs are, naturally, added or multiplied mod a; the second components are added or multiplied mod b, so we speak of pairs being congruent (mod a, mod b).

Example. Adding and multiplying mod 15.

Consider 8 and 9, and their sum and product mod 15. We have

$$8 \text{ represented by } (2, 3)$$
$$9 \text{ represented by } (0, 4).$$

Adding these pairs mod 3 in the first component and mod 5 in the second, we get

$$(2, 3) + (0, 4) = (2 + 0, 3 + 4) = (2, 7)$$
$$\equiv (2, 2) \quad (\bmod 3, \bmod 5).$$

(2,2) is the pair that represents 2, and indeed $8 + 9 \equiv 2 \pmod{15}$.

Similarly, if we multiply $(2,3)$ and $(0,4)$, mod 3 in the first component and mod 5 in the second, we get

$$(2,3) \times (0,4) = (2 \times 0, 3 \times 4) = (0,12)$$
$$\equiv (0,2) \quad (\text{mod } 3, \text{mod } 5).$$

$(0,2)$ is the pair that represents 12, and indeed $8 \times 9 \equiv 12 \ (\text{mod } 15)$.

The **full Chinese remainder theorem** says that the pair $(m \bmod a, m \bmod b)$ corresponds to m, mod ab, and that

$$(m \bmod a, m \bmod b) + (n \bmod a, n \bmod b)$$
$$\equiv (m+n \bmod a, m+n \bmod b) \quad (\text{mod } a, \text{mod } b)$$

and

$$(m \bmod a, m \bmod b) \times (n \bmod a, n \bmod b)$$
$$\equiv (mn \bmod a, mn \bmod b) \quad (\text{mod } a, \text{mod } b)$$

This follows easily from addition and multiplication of congruences.

$$m \bmod a \ \text{ is } \equiv m \quad (\text{mod } a),$$
$$n \bmod a \ \text{ is } \equiv n \quad (\text{mod } a),$$

and therefore, by addition of congruences,

$$(m \bmod a) + (n \bmod a) \ \text{ is } \equiv m+n \quad (\text{mod } a).$$

Similarly for addition mod b, and for multiplication mod a and mod b. $\square$

This version of the theorem shows that the pairs $(n \bmod a, n \bmod b)$ not only *correspond* 1-to-1 to numbers $n \ (\text{mod } ab)$ (we need $\gcd(a,b) = 1$ for this part), but also *behave the same under* $+$ *and* $\times$ (mod a, mod b).

Invertible elements

The modern Chinese remainder theorem gives a very clear picture of the group $(\mathbb{Z}/ab\mathbb{Z})^{\times}$ of invertible elements under multiplication mod ab.

As we have just seen, when $\gcd(a,b) = 1$, n behaves (mod ab) as the pair $(n \bmod a, n \bmod b)$ does (mod a, mod b). In particular, *n has an inverse*, (mod ab), *if and only if n mod a has an inverse* (mod a) *and n mod b has an inverse* (mod b).

Example. Invertible elements, mod 15.

These are the n for which the remainder pairs $(n \bmod 3, n \bmod 5)$ have inverses (mod 3, mod 5). Since 3 and 5 are primes, these are precisely the pairs in which $n \bmod 3$ and $n \bmod 5$ are *nonzero*.

There are two nonzero elements mod 3 (namely 1 and 2) and four nonzero elements mod 5 (namely 1, 2, 3 and 4), hence there are

$$2 \times 4 = 8$$

pairs $(n \bmod 3, n \bmod 4)$ of nonzero elements, and hence eight invertible elements n (mod 15). They can be read off the table in Section 9.6 as the numbers 1, 2, 4, 7, 8, 11, 13 ,14—those for which the corresponding pairs have no zeros.

This example generalizes to a key theorem about the φ function.

Multiplicative property of φ. *When* $\gcd(a,b) = 1$, $\varphi(ab) = \varphi(a)\varphi(b)$.

Proof. By the criterion for inverses in Section 3.6, there are $\varphi(a)$ invertible elements (mod a) and $\varphi(b)$ invertible elements (mod b). Therefore, if $\gcd(a,b) = 1$ there are $\varphi(a)\varphi(b)$ invertible pairs $(n \bmod a, n \bmod b)$; that is, $\varphi(a)\varphi(b)$ invertible elements (mod ab). But the number of invertible elements (mod ab) is $\varphi(ab)$. Hence *if* $\gcd(a,b) = 1$, *we have*

$$\varphi(ab) = \varphi(a)\varphi(b) \qquad\qquad \square$$

Exercises

Thanks to the multiplicative property, we can now complete our search for an explicit formula for $\varphi(n)$, begun in the exercises to Section 3.6.

9.7.1 Using Exercise 3.6.3, show that $\varphi(n) = n(1 - \frac{1}{p_1}) \cdots (1 - \frac{1}{p_k})$, where p_1, p_2, $\dots, p_k$ are the distinct prime divisors of n.

9.7.2 Use the formula to show that $\varphi(60) = 16$.

9.8   Proof of quadratic reciprocity

A formula for $\left(\frac{p}{q}\right)$ **and** $\left(\frac{q}{p}\right)$

We now give a formula that simultaneously exhibits $\left(\frac{p}{q}\right)$ and $\left(\frac{q}{p}\right)$ in a product of pairs (mod p, mod q). This formula is used to prove quadratic reciprocity below, following the argument of Rousseau (1991).

When p and q are different odd primes we consider the invertible numbers mod pq, which are those divisible by neither p nor q. The invertible x in the range $1 \leq x \leq \frac{pq-1}{2}$, taken mod p, consist of $\frac{q-1}{2}$ sequences $1, 2, \ldots, p-1$ and the "half sequence" $1, 2, \ldots, \frac{p-1}{2}$, minus the $\frac{p-1}{2}$ multiples $q, 2q, \ldots \frac{p-1}{2}q$ of q in this range.

The mod p product of these invertible x is therefore

$$\overset{\text{invertible } x}{\underset{1 \leq x \leq \frac{pq-1}{2}}{\prod}} x \equiv (p-1)!^{\frac{q-1}{2}} \left(\frac{p-1}{2}\right)! \bigg/ q^{\frac{p-1}{2}} \left(\frac{p-1}{2}\right)! \quad (\text{mod } p)$$

$$\equiv (-1)^{\frac{q-1}{2}} \left(\frac{q}{p}\right) \quad (\text{mod } p),$$

since $\left(\frac{p-1}{2}\right)!$ cancels, $(p-1)! \equiv -1$ by Wilson's theorem, and $q^{\frac{p-1}{2}} \equiv \left(\frac{q}{p}\right)$ by Euler's criterion. Similarly, the mod q product of these invertible x is

$$\overset{\text{invertible } x}{\underset{1 \leq x \leq \frac{pq-1}{2}}{\prod}} x \equiv (-1)^{\frac{p-1}{2}} \left(\frac{p}{q}\right) \quad (\text{mod } q),$$

and therefore the (mod p, mod q) product of pairs (x,x) is

$$\overset{\text{invertible } x}{\underset{1 \leq x \leq \frac{pq-1}{2}}{\prod}} (x,x) \equiv \left((-1)^{\frac{q-1}{2}} \left(\frac{q}{p}\right), (-1)^{\frac{p-1}{2}} \left(\frac{p}{q}\right) \right) \quad (\text{mod } p, \text{mod } q) \quad (1)$$

Completion of the proof

Now we evaluate $\prod(x,x)$ over the invertible x in a different way, expressing it in powers of -1 alone. Using the Chinese remainder theorem, we view it as a product of pairs (a,b), with a and b varying independently over suitable ranges.

The x in the range $1 \leq x \leq \frac{pq-1}{2}$ include exactly one number from each pair $\{x, -x\}$ (mod pq) with $1 \leq x \leq pq-1$. Hence the corresponding remainder pairs,

$$(a,b) = (x \bmod p, x \bmod q),$$

include exactly one of each pair $\{(a,b), (-a,-b)\}$. We get exactly one of each (mod p, mod q) by taking the ranges $1 \leq a \leq p-1$ and $1 \leq b \leq \frac{q-1}{2}$.

This makes the sign uncertain, but at any rate

$$\prod_{1 \le x \le \frac{pq-1}{2}}^{\text{invertible } x} (x,x) \equiv \pm \left((p-1)!^{\frac{q-1}{2}}, ((q-1)/2)!^{p-1} \right) \quad (\bmod\ p, \bmod\ q),$$

(2)

since each value of a, $1 \le a \le p-1$, occurs in $(q-1)/2$ pairs, and each value of b, $1 \le b \le (q-1)/2$, occurs in $p-1$ pairs. Thanks to Wilson's theorem, we can express the powers of factorials in (2) as powers of -1.

Since $(p-1)! \equiv -1 \ (\bmod\ p)$, the first component $\equiv (-1)^{\frac{q-1}{2}} \ (\bmod\ p)$. To find $((q-1)/2)! \ (\bmod\ q)$ we shape $-1 \equiv (q-1)!$ as in Section 6.5:

$$-1 \equiv (q-1)! \quad (\bmod\ q)$$
$$\equiv 1 \times 2 \times \cdots \times ((q-1)/2)$$
$$\times (-(q-1)/2) \times \cdots \times (-2) \times (-1) \quad (\bmod\ q)$$
$$\equiv ((q-1)/2)!^2 (-1)^{\frac{q-1}{2}} \quad (\bmod\ q).$$

Therefore

$$((q-1)/2)!^2 \equiv (-1)(-1)^{\frac{q-1}{2}} \quad (\bmod\ q).$$

Raising both sides to the power $\frac{p-1}{2}$, we get the second component of (2),

$$((q-1)/2)!^{p-1} \equiv (-1)^{\frac{p-1}{2}} (-1)^{\frac{p-1}{2}\frac{q-1}{2}} \quad (\bmod\ q).$$

Thus the expression (2) for $\prod(x,x) \ (\bmod\ p, \bmod\ q)$, reduces to

$$\prod_{1 \le x \le \frac{pq-1}{2}}^{\text{invertible } x} (x,x) \equiv \pm \left((-1)^{\frac{q-1}{2}}, (-1)^{\frac{p-1}{2}} (-1)^{\frac{p-1}{2}\frac{q-1}{2}} \right) \quad (\bmod\ p, \bmod\ q)$$

(3)

Equating (3) with (1) in the previous subsection we get either

$$\left(\frac{q}{p} \right) = 1 \quad \text{and} \quad \left(\frac{p}{q} \right) = (-1)^{\frac{p-1}{2}\frac{q-1}{2}}$$

or

$$\left(\frac{q}{p} \right) = -1 \quad \text{and} \quad \left(\frac{p}{q} \right) = -(-1)^{\frac{p-1}{2}\frac{q-1}{2}}.$$

In either case,

$$\left(\frac{p}{q} \right) \left(\frac{q}{p} \right) = (-1)^{\frac{p-1}{2}\frac{q-1}{2}}. \qquad \square$$

Exercises

With the help of quadratic reciprocity we can find, for any *fixed* odd prime p, the primes q for which p is a square mod q. They fall into a finite number of arithmetic progressions (as we have already seen they do for $p = -1$ and $p = 2$). Here is what happens for $p = 3$.

9.8.1 Explain why every odd prime q is of one of the forms $12n + 1$, $12n + 5$, $12n + 7$, or $12n + 11$.

9.8.2 Use quadratic reciprocity with remaindering, as in Section 9.5, to show that 3 is a square mod q for precisely the odd primes q of the form $12n + 1$ and $12n + 11$.

By multiplying the values of $\left(\frac{3}{q}\right)$ found in Exercise 9.8.2 by the corresponding values of $\left(\frac{-1}{q}\right)$ we can also obtain the values of $\left(\frac{-3}{q}\right)$:

9.8.3 Show that -1 is a square mod q for the odd primes q of the form $12n + 1$ and $12n + 5$, and a nonsquare for those of the form $12n + 7$ and $12n + 11$. Deduce that -3 is a square mod q for the odd primes q of the form $12n + 1$ and $12n + 7$.

Similarly, we can find the values of $\left(\frac{5}{q}\right)$ and $\left(\frac{-5}{q}\right)$:

9.8.4 Show that the odd primes q for which 5 is a square mod q are precisely those of the form $20n + 1$, $20n + 9$, $20n + 11$, and $20n + 19$. Test this result by showing that 5 is a square mod 41, 29, 11, and 19.

9.8.5 Show that the odd primes q for which -5 is a square mod q are precisely those of the form $20n + 1$, $20n + 3$, $20n + 7$, and $20n + 9$. Test this result by showing that -5 is a square mod 41, 23, 7, and 29.

9.9   Discussion

As we have suggested over the last few chapters, quadratic reciprocity emerged from the study of primes represented by quadratic forms such as $x^2 + y^2$, $x^2 + 2y^2$, and $x^2 + 3y^2$. Fermat was the first to raise and answer such questions, but his methods are not known. As far as we know, Euler was the first to recognize the role of quadratic reciprocity and to prove it in special cases.

The first to attempt a general proof was Legendre (1785). However, his proof depended on the unproved assumption that any arithmetic progression $an + b$ with $\gcd(a, b) = 1$ contains infinitely many primes. This assumption is easily proved in certain cases (such as the cases $4n + 1$ and

$4n + 3$ mentioned in the exercises to Section 6.3 and in Section 6.7) but the general theorem is harder to prove than reciprocity itself. The first proof was given by Dirichlet (1837), and the deep analytic methods he devised to prove it are still the standard approach to primes in arithmetic progressions.

Gauss found the first proof of quadratic reciprocity on April 18, 1796, when he was not quite 19. It is a long and ugly proof, and by the time he published it in his *Disquisitiones Arithmeticae* of (1801) he had found two more proofs; one using quadratic forms and the other using roots of unity. Quadratic reciprocity was Gauss's favorite theorem and altogether he gave eight proofs of it. Since then, many other mathematicians have published proofs—some of them variations or simplifications of Gauss, and others introducing new ideas.

Like Pythagoras' theorem in geometry, quadratic reciprocity is a core theorem in number theory, bound to arise no matter how one approaches quadratic Diophantine equations. This is why the theorem has so many proofs: all roads lead to it, and each road shows it from a different angle. A comprehensive history of quadratic reciprocity, including a table and classification of 196 (!) proofs given up to the year 2000, may be found in Lemmermeyer (2000). Another book of interest is Pieper (1978), which discusses 14 different proofs in detail.

The law of quadratic reciprocity generalizes to cubic, biquadratic, and higher power reciprocity laws. Just as the quadratic character has values ± 1 (the square roots of 1), there is a *cubic character* with values 1, ζ_3, ζ_3^2 (the cube roots of 1), and a *biquadratic character* with values ± 1, $\pm i$ (the fourth roots of 1). These generalizations were not made because mathematicians had run out of things to say about quadratic forms—quite the contrary; quadratic forms themselves demand that cubes and fourth powers be considered. This was discovered by Euler, who noticed the following results (later proved by Gauss):

p is a prime $x^2 + 27y^2 \Leftrightarrow p = 3n + 1$ and 2 is a cube mod p

p is a prime $x^2 + 64y^2 \Leftrightarrow p = 4n + 1$ and 2 is a fourth power mod p

The cubic reciprocity law was found by Eisenstein (1844) and it requires investigation of $\mathbb{Z}[\zeta_3]$, which is why we call $\mathbb{Z}[\zeta_3]$ the Eisenstein integers. (Cubic reciprocity was already known to Gauss, but he did not publish his results.) Likewise, biquadratic reciprocity was discovered by Gauss and it requires investigation of $\mathbb{Z}[i]$. This in fact was the purpose of Gauss (1832), where the basic properties of $\mathbb{Z}[i]$ were first published.

An nth power reciprocity law similarly involves the cyclotomic integers $\mathbb{Z}[\zeta_n]$, with all their attendant difficulties such as the failure of unique prime factorization discovered by Kummer (1844). In the case of nth power reciprocity (and unlike the case of Fermat's last theorem), Kummer overcame these difficulties completely with his theory of ideal numbers, and published an nth power reciprocity law in 1850. Eisenstein, also using Kummer's theory of ideal numbers, published a different version of nth power reciprocity in the same year. A modern proof of Eisenstein's reciprocity law may be found in Ireland and Rosen (1982), pp. 215–218, and the history of all reciprocity laws up to 1850 may be found in Lemmermeyer (2000).

10

Rings

PREVIEW

This chapter unites many of the algebraic structures encountered in this book—the integers, the integers mod n, and the various extensions of the integer concept by Gauss, Eisenstein and Hurwitz—in the single abstract concept of *ring*.

We begin with the general ring concept, specified by certain axioms for $+$ and $\times$, and observe how these axioms suffice to capture general concepts of divisibility, primes, and units. The concept of *field*—a ring in which all nonzero elements are units—is briefly discussed, and the main examples $\mathbb{Q}$, $\mathbb{R}$, $\mathbb{C}$, and $\mathbb{Z}/p\mathbb{Z}$ are reviewed.

We then specialize to rings of algebraic integers, and particularly quadratic integers. We define algebraic numbers and algebraic integers and use Dedekind's linear algebra approach to show that the algebraic integers are closed under $+$, $-$, and $\times$, and hence form a ring.

The special case of quadratic integer rings, and the quadratic fields $\mathbb{Q}(\sqrt{d})$ that contain them, is examined in more detail. We give a general explanation of the phenomenon of quadratic integers, such as $\frac{-1+\sqrt{-3}}{2}$, that "look fractional", by determining the integers of all the fields $\mathbb{Q}(\sqrt{d})$ for $d \in \mathbb{Z}$.

The concept of norm, previously seen in special cases, is given a uniform definition over all quadratic fields. Finally, we specialize further to the *imaginary* quadratic fields—the $\mathbb{Q}[\sqrt{d}]$ for negative integers d. These enjoy somewhat simpler properties than the $\mathbb{Q}(\sqrt{d})$ for positive d. For example, the integers of an imaginary quadratic field include only finitely many units (at most six).

10.1   The ring axioms

The integers $\mathbb{Z}$, with their operations of $+$ and $\times$, are the first ring studied in mathematics, and some basic properties of $\mathbb{Z}$ are taken as the defining properties (axioms) of rings in general. We briefly mentioned them in Section 1.3, in connection with abelian groups. An explicit list of the axioms is as follows. For all integers a, b and c we have:

$$a + (b+c) = (a+b) + c \qquad \text{(associative law)}$$
$$a + b = b + a \qquad \text{(commutative law)}$$
$$a + (-a) = 0 \qquad \text{(additive inverse property)}$$
$$a + 0 = a \qquad \text{(identity property of 0)}$$

There is a similar set of rules describing the behavior of $\times$.

$$a \times (b \times c) = (a \times b) \times c \qquad \text{(associative law)}$$
$$a \times b = b \times a \qquad \text{(commutative law)}$$
$$a \times 1 = a \qquad \text{(identity property of 1)}$$
$$a \times 0 = 0 \qquad \text{(property of 0)}$$

and finally, there is a rule for the interaction of $+$ and $\times$:

$$a \times (b+c) = a \times b + a \times c \qquad \text{(distributive law)}$$

Strictly speaking, these are the defining properties of a *commutative* ring. We have also dealt occasionally with *noncommutative* rings, such as the quaternions $\mathbb{H}$, which satisfy all the above axioms except the commutative law for $\times$. The quaternions, however, are close to number rings in having a multiplicative norm, and noncommutative ring theory in general has a rather different flavor.

It is fair to say that most of ring theory deals with objects that behave like integers, and the example of $\mathbb{Z}$ helps us to anticipate which concepts will be relevant and helpful in dealing with unfamiliar, but "integer-like" objects. Indeed, we have already used the word "integer" for extensions of $\mathbb{Z}$ such as $\mathbb{Z}[i]$ (the Gaussian integers), $\mathbb{Z}[\zeta_3]$ (the Eisenstein integers), and $\mathbb{Z}[\frac{1+i+j+k}{2}, i, j, k]$ (the Hurwitz integers).

Divisibility and primes

In the examples of generalized "integers" we have seen the importance of the ordinary integer concepts of divisibility and primes. In any ring, we say that *b divides a* if there is a *c* such that

$$a = bc.$$

Other ways to say this are that *a* is *divisible by b*, that *b* is a *divisor* of *a*, or that *a* is a *multiple* of *b*. Divisibility is an interesting concept in $\mathbb{Z}$ because an arbitrary integer *a* is generally *not* divisible by another integer *b*. In fact, it can be hard to decide whether *a* has any divisors at all, apart the obvious ones ± 1 and $\pm a$. An ordinary integer with no divisors except the obvious ones is called *prime*. In an arbitrary ring *R* the concept of prime is the same, except that the place of ± 1 is taken by the *units* of *R*: the elements of *R* that divide 1 (or, equivalently, the invertible elements of *R*). Thus we call $a \in R$ *prime* if *a* is divisible only by units and units times *a* (the latter are called *associates* of *a*).

   Even in $\mathbb{Z}$ the primes form no clear pattern, and of course prime numbers figure in many of the classic unsolved problems about $\mathbb{Z}$. Thus the development of ring theory has been heavily influenced by the problem of understanding primes. The best understood rings, such as $\mathbb{Z}[i]$, tend to be those whose primes behave like the primes in $\mathbb{Z}$. In some rings where this is not the case—specifically, where unique prime factorization fails—it has been found worthwhile to create "ideal" primes that behave better than the actual primes. We take up the story of these "ideal" primes in Chapters 11 and 12.

Exercises

Recall from Chapter 8 that the quaternions $\mathbb{H}$ are certain 2×2 matrices. In fact, the (noncommutative) ring concept extends much further in this direction, to rings of $n \times n$ matrices for any fixed natural number *n*. For the moment we consider all such matrices with complex number entries, with zero matrix 0 and identity matrix 1.

10.1.1 Check that the axioms for $+$ hold.

10.1.2 Check that the axioms for $\times$ hold, except the commutative law (when $n \geq 2$). You need *not* use explicit multiplication to prove the associative law. Why?

10.1.3 Check that the distributive law holds.

10.2   Rings and fields

The sets $\mathbb{Q}$ (rational numbers), $\mathbb{R}$ (real numbers), and $\mathbb{C}$ (complex numbers) are also rings because they obviously have the properties of $+$, $-$, and $\times$ listed in the previous section. This is no surprise because $\mathbb{Q}$, $\mathbb{R}$, and $\mathbb{C}$ were always intended to *extend* the concept of integer, retaining all the ring properties and adding more. One thing that $\mathbb{Q}$, $\mathbb{R}$, and $\mathbb{C}$ have and $\mathbb{Z}$ does not is a *multiplicative inverse* a^{-1} for each nonzero element, with the characteristic property that $aa^{-1} = 1$. A commutative ring with a multiplicative inverse for each nonzero element is called a *field*.

Fields are not really typical rings, and their theory has quite a different flavor from ring theory. In particular, divisibility is not an interesting notion in a field because a nonzero element b divides *any* element a (with quotient $a/b = ab^{-1}$). Likewise, the concept of unit is not interesting because all nonzero elements of the field are units. Nevertheless, fields play an important role in ring theory. Many of the rings we used in earlier chapters, such as $\mathbb{Z}[i]$, $\mathbb{Z}[\sqrt{-2}]$, and $\mathbb{Z}[\zeta_3]$, are embedded in the complex field $\mathbb{C}$. Since $\mathbb{C}$ has all the ring properties (associative laws, commutative laws, and so on) the same will be true of any subset $R \subseteq \mathbb{C}$ for which the expressions $a + b$, $-a$, $a \times b$ mentioned in the ring axioms make sense. That is, a set $R \subseteq \mathbb{C}$ is a ring provided R is *closed* under $+$, $-$, and $\times$. This means that if $a, b \in R$ then $a + b, a - b, a \times b \in R$.

The rings just mentioned exemplify the process of "closing" a set under the operations $+$, $-$, and $\times$. In these examples, we take a number a not in $\mathbb{Z}$ and close the set $\mathbb{Z} \cup \{a\}$ by forming all possible sums, differences and products involving a and the integers. The result is called the ring $\mathbb{Z}[a]$.

If we then take an element b not in $\mathbb{Z}[a]$ and repeat the process of closing under $+$, $-$, and $\times$, the resulting ring is called $\mathbb{Z}[a, b]$, and so on. This is in accordance with the notations we have already used for the rings $\mathbb{Z}[\mathbf{i}, \mathbf{j}, \mathbf{k}]$ and $\mathbb{Z}[\frac{1+\mathbf{i}+\mathbf{j}+\mathbf{k}}{2}, \mathbf{i}, \mathbf{j}, \mathbf{k}]$ of quaternion integers. Any subset of the quaternions closed under $+$, $-$, and $\times$ is a ring, though generally noncommutative. (Indeed, it is clear that any subring of $\mathbb{H}$ containing nonzero multiples of more than one of $\mathbf{i}, \mathbf{j}, \mathbf{k}$ is noncommutative.)

When the ring is a subset of $\mathbb{C}$, we can form its closure under division by nonzero elements, and obtain a field. If we take an element a not in such a field F, and again close under $+$, $-$, $\times$, and $\div$ (by nonzero elements), then the result is called $F(a)$. We occasionally use this round bracket notation for fields such as $\mathbb{Q}(\sqrt{2})$, though in fact here we have $\mathbb{Q}[\sqrt{2}] = \mathbb{Q}(\sqrt{2})$.

Finite rings and fields

Finite rings came up implicitly in Section 3.2, where we introduced the notation $\mathbb{Z}/n\mathbb{Z}$ for the set of congruence classes mod n under the operations of $+$ and $\times$ for congruence classes. We did not comment on it at the time, but it is easy to see that $\mathbb{Z}/n\mathbb{Z}$ is a ring. Its ring properties are "inherited" from its parent ring $\mathbb{Z}$. For example, the $+$ operation on congruence classes is commutative because

$$
\begin{aligned}
(n\mathbb{Z}+a)+(n\mathbb{Z}+b) &= n\mathbb{Z}+(a+b) \quad \text{by definition of } + \text{ in } \mathbb{Z}/n\mathbb{Z} \\
&= n\mathbb{Z}+(b+a) \quad \text{by commutative law for } + \text{ in } \mathbb{Z} \\
&= (n\mathbb{Z}+b)+(n\mathbb{Z}+a) \quad \text{by definition of } + \text{ in } \mathbb{Z}/n\mathbb{Z}
\end{aligned}
$$

We also showed, in Section 3.3, that every nonzero element of $\mathbb{Z}/p\mathbb{Z}$ has a multiplicative inverse when p is prime. Thus the finite ring $\mathbb{Z}/p\mathbb{Z}$ is a field.

The units of the ring $\mathbb{Z}/n\mathbb{Z}$ are particularly interesting, since they form the group $(\mathbb{Z}/n\mathbb{Z})^{\times}$, which can be quite complicated. It is easy to see (exercises) that the units of any ring form a group. If the ring is noncommutative then the group of units may be noncommutative, as we saw in the case of the Hurwitz integers in the exercises to Section 8.5. We have also seen that infinite groups of units occur, for example in $\mathbb{Z}[\sqrt{2}]$. This was implicit in Section 5.4.

Exercises

10.2.1 Show that the product of two units in a ring R is also a unit of R.

10.2.2 Show also that the multiplicative inverse of a unit is a unit, and hence that the units of any ring form a group.

The ring $\mathbb{Z}/n\mathbb{Z}$, for n not prime, differs from $\mathbb{Z}$ in having *zero divisors*— nonzero elements whose product is zero.

10.2.3 Give an example of a zero divisor in $\mathbb{Z}/4\mathbb{Z}$.

10.2.4 Explain why $\mathbb{Z}/n\mathbb{Z}$ has zero divisors for any n that is not prime.

Zero divisors prevent us from extending $\mathbb{Z}/n\mathbb{Z}$ to a field by adjoining "fractions", the way we extend $\mathbb{Z}$ to $\mathbb{Q}$ for example.

10.2.5 If a is a zero divisor in $\mathbb{Z}/n\mathbb{Z}$, show that we cannot consistently adjoin an element a^{-1} such that $aa^{-1} = 1$ in $\mathbb{Z}/n\mathbb{Z}$.

10.3   Algebraic integers

Algebraic numbers

Many concepts of ring theory originated in Dedekind's theory of algebraic integers. Dedekind generalized the idea of embedding the ordinary integers in the field of rational numbers by embedding various rings of *algebraic* integers in fields of *algebraic* numbers.

Definition. A number $\alpha \in \mathbb{C}$ is *algebraic* if

$$a_m \alpha^m + a_{m-1} \alpha^{m-1} + \cdots + a_1 \alpha + a_0 = 0 \quad \text{where} \quad a_0, a_1, \ldots, a_m \in \mathbb{Z},$$

and it is of *degree m* if it satisfies no such equation of lower degree.

Examples are:

- rational numbers, which are the algebraic numbers of degree 1,

- $\sqrt{2}$, which is of degree 2 because it satisfies $x^2 - 2 = 0$ but no equation of lower degree (since $\sqrt{2}$ is irrational).

Like the rationals, the set of all algebraic numbers is a field, though this is not obvious. It is not even clear that the algebraic numbers are closed under $+$, for example,

$$\sqrt[3]{2} \text{ satisfies } x^3 - 2 = 0, \text{ and hence is algebraic,}$$
$$\sqrt[5]{3} \text{ satisfies } x^5 - 3 = 0, \text{ and hence is algebraic,}$$
$$\text{but what equation does } \sqrt[3]{2} + \sqrt[5]{3} \text{ satisfy?}$$

We do not prove that the algebraic numbers form a field here, but we can show that $\sqrt[3]{2} + \sqrt[5]{3}$ satisfies a polynomial equation with integer coefficients. This follows from what we are about to prove about algebraic integers, whose definition we recall from Section 7.4.

Definition. A number $\alpha \in \mathbb{C}$ is an *algebraic integer* if it satisfies a *monic* polynomial equation with integer coefficients, that is

$$\alpha^m + a_{m-1} \alpha^{m-1} + \cdots + a_1 \alpha + a_0 = 0 \quad \text{where} \quad a_0, a_1, \ldots, a_{m-1} \in \mathbb{Z}.$$

Examples are:

- ordinary integers, which are algebraic integers of degree 1,

- $\sqrt[3]{2}$ and $\sqrt[5]{3}$, because they satisfy the monic equations $x^3 - 2 = 0$ and $x^5 - 3 = 0$ respectively,

- $(-1 + \sqrt{-3})/2$, because it satisfies $x^2 + x + 1 = 0$.

On the other hand, the only rational algebraic integers are the ordinary integers, as we proved in Section 7.4.

The ring of algebraic integers

The algebraic integers lie in the ring $\mathbb{C}$. Hence to prove that they form a ring it suffices to prove that they are closed under $+$, $-$, and $\times$. This was first proved by Eisenstein, but we follow a more modern proof given by Dedekind.

Closure properties of algebraic integers. *If α and β are algebraic integers then so are $\alpha + \beta$, $\alpha - \beta$, and $\alpha\beta$.*

Proof. By hypothesis, α and β satisfy equations

$$\alpha^m + a_{m-1}\alpha^{m-1} + \cdots + a_1\alpha + a_0 = 0 \quad \text{where} \quad a_0, a_1, \ldots, a_{m-1} \in \mathbb{Z},$$
$$\beta^n + b_{n-1}\beta^{n-1} + \cdots + b_1\beta + b_0 = 0 \quad \text{where} \quad b_0, b_1, \ldots, b_{n-1} \in \mathbb{Z}.$$

These equations show:

- $\alpha^m = -a_0 - a_1\alpha - \cdots - a_{m-1}\alpha^{m-1}$ is a linear combination of 1, α, ..., α^{m-1} with integer coefficients.

- $\alpha^{m+1} = -a_0\alpha - a_1\alpha^2 - \ldots - a_{m-1}\alpha^m$ is a linear combination of $1, \alpha, \ldots, \alpha^{m-1}$ with integer coefficients (since α^m can be rewritten in terms of $1, \alpha, \ldots, \alpha^{m-1}$).

- Similarly, every power of α is a linear combination of $1, \alpha, \ldots, \alpha^{m-1}$ with integer coefficients.

- Likewise, every power of β is a linear combination of $1, \beta, \ldots, \beta^{n-1}$ with integer coefficients.

- Therefore, every polynomial in α and β is a linear combination of terms $\alpha^i \beta^j$ with $0 \le i \le m-1$ and $0 \le j \le n-1$.

Thus if we denote the mn products $\alpha^i \beta^j$ by $\omega_1, \omega_2, \ldots, \omega_{mn}$, we can write any polynomial ω in α, β with integer coefficients as a linear combination of $\omega_1, \omega_2, \ldots, \omega_{mn}$ with integer coefficients. In particular, if ω is any one of $\alpha + \beta$, $\alpha - \beta$, or $\alpha\beta$ we have

$$\omega = k_1\omega_1 + \cdots + k_{mn}\omega_{mn} \quad \text{for some} \quad k_1, k_2, \ldots, k_{mn} \in \mathbb{Z}. \quad (*)$$

From this we obtain mn equations in the mn "unknowns" $\omega_1, \omega_2, \ldots, \omega_{mn}$ by multiplying (*) by $\omega_1, \omega_2, \ldots, \omega_{mn}$ and rewriting each right-hand side as a linear combination of $\omega_1, \omega_2, \ldots, \omega_{mn}$ with integer coefficients $k_s^{(t)}$:

$$\omega\omega_1 = k_1'\omega_1 + \cdots + k_{mn}'\omega_{mn}$$
$$\omega\omega_2 = k_1''\omega_1 + \cdots + k_{mn}''\omega_{mn}$$
$$\vdots$$
$$\omega\omega_{mn} = k_1^{(mn)}\omega_1 + \cdots + k_{mn}^{(mn)}\omega_{mn}$$

These are homogeneous equations in $\omega_1, \omega_2, \ldots, \omega_{mn}$ with a nonzero solution, and hence their determinant must be zero. That is,

$$\begin{vmatrix} k_1' - \omega & k_2' & \cdots & k_{mn}' \\ k_1'' & k_2'' - \omega & \cdots & k_{mn}'' \\ \vdots & & & \vdots \\ k_1^{(mn)} & k_2^{(mn)} & \cdots & k_{mn}^{(mn)} - \omega \end{vmatrix} = 0.$$

This determinant is a polynomial in ω, with coefficients in $\mathbb{Z}$, and with coefficient of ω^{mn} equal to ± 1. Hence $\omega = \alpha + \beta, \alpha - \beta$, or $\alpha\beta$ is an algebraic integer. □

Exercises

The units among the algebraic integers divide 1, by definition, and hence they are the algebraic integers α such that α^{-1} is also an algebraic integer.

10.3.1 Deduce that α is an algebraic integer unit if and only if α satisfies a monic polynomial equation with integer coefficients and constant term ± 1.

10.3.2 Verify that such polynomials are satisfied by the units $\pm\zeta_3$ and $\pm\zeta_3^2$ of $\mathbb{Z}[\zeta_3]$.

10.4   Quadratic fields and their integers

The ring of all algebraic integers has some inconvenient properties. For example, the square root of any algebraic integer α is also an algebraic integer, and hence α has the factorization $\alpha = \sqrt{\alpha}\sqrt{\alpha}$. This shows that there are no "primes" in the ring of algebraic integers. For this reason, one usually works in a ring of algebraic integers of bounded degree, such as the rings $\mathbb{Z}[i]$ and $\mathbb{Z}[\sqrt{-2}]$ we used previously. We now generalize the latter examples to the *ring of integers of a quadratic field*.

Each quadratic field can be written $\mathbb{Q}(\sqrt{d})$, where $d \in \mathbb{Z}$ and $\mathbb{Q}(\sqrt{d})$ is "the smallest field containing $\mathbb{Q}$ and $\sqrt{d}$" or, in other words, the result of closing the set $\mathbb{Q} \cup \{\sqrt{d}\}$ under the operations $+$, $-$, $\times$, and $\div$ (by nonzero members). It is closure under $\div$, as well as $+$, $-$, and $\times$, that produces a field, and we use the round bracket notation to distinguish the result from closure under $+$, $-$, and $\times$ alone, which produces a ring but not necessarily a field. (For example, $\mathbb{Z}[i]$ is the closure of $\mathbb{Z} \cup \{i\}$ under $+$, $-$, and $\times$, and it is not a field.)

The round bracket notation is superfluous in this case, because in fact $\mathbb{Q}(\sqrt{d}) = \mathbb{Q}[\sqrt{d}]$.

Characterization of $\mathbb{Q}(\sqrt{d})$. $\mathbb{Q}(\sqrt{d}) = \mathbb{Q}[\sqrt{d}] = \{a + b\sqrt{d} : a, b \in \mathbb{Q}\}$.

Proof. Each number $a + b\sqrt{d}$ with $a, b \in \mathbb{Q}$ certainly results from $\sqrt{d}$ and the members a, b of $\mathbb{Q}$ by $+$ and $\times$. Hence

$$\{a + b\sqrt{d} : a, b \in \mathbb{Q}\} \subseteq \mathbb{Q}[\sqrt{d}] \subseteq \mathbb{Q}(\sqrt{d}).$$

Conversely, we can show that $\{a + b\sqrt{d} : a, b \in \mathbb{Q}\}$ is closed under $+$, $-$, $\times$, and hence $\{a + b\sqrt{d} : a, b \in \mathbb{Q}\} \supseteq \mathbb{Q}[\sqrt{d}]$. We also show it is closed under $\div$, hence $\{a + b\sqrt{d} : a, b \in \mathbb{Q}\} \supseteq \mathbb{Q}(\sqrt{d})$.

The set $\{a + b\sqrt{d} : a, b \in \mathbb{Q}\}$ is obviously closed under $+$ and $-$. It is closed under $\times$ because

$$(a_1 + b_2\sqrt{d})(a_2 + b_2\sqrt{d}) = a_1 a_2 + b_1 b_2 d + (a_1 b_2 + a_2 b_1)\sqrt{d}.$$

And it is closed under ÷ (by nonzero members) because

$$\frac{1}{a+b\sqrt{d}} = \frac{a-b\sqrt{d}}{(a+b\sqrt{d})(a-b\sqrt{d})}$$

$$= \frac{a-b\sqrt{d}}{a^2-b^2 d}$$

$$= \frac{a}{a^2-b^2 d} - \frac{b}{a^2-b^2 d}\sqrt{d}. \qquad \square$$

The integers of $\mathbb{Q}(\sqrt{d})$ include $\pm\sqrt{d}$ (since $\pm\sqrt{d}$ satisfy the monic equation $x^2 - d = 0$) and hence they include all members of $\mathbb{Z}[\sqrt{d}]$ (by closure of the algebraic integers under +). But sometimes they include more, which is why we have to use the awkward phrase "integers of $\mathbb{Q}(\sqrt{d})$" instead of $\mathbb{Z}[\sqrt{d}]$. For example, $\mathbb{Q}(\sqrt{-3})$ includes $(-1+\sqrt{-3})/2$, and the latter is an algebraic integer because it satisfies $x^2 + x + 1 = 0$.

The precise situation is described in the following theorem. Before we state the theorem and begin its proof we note that the integer d in $\mathbb{Q}(\sqrt{d})$ can be assumed *squarefree*, that is, not divisible by any square > 1. This is so because if $d = n^2 c$ for some $n, c \in \mathbb{Z}$, then $\mathbb{Q}(\sqrt{d}) = \mathbb{Q}(n\sqrt{c})$, which equals $\mathbb{Q}(\sqrt{c})$ by closure under × and ÷. The other thing to remember is that any square is $\equiv 0$ or 1 (mod 4).

Integers of $\mathbb{Q}(\sqrt{d})$. When $d \not\equiv 1$ (mod 4) *the integers of* $\mathbb{Q}(\sqrt{d})$ *are the* $a+b\sqrt{d}$ *with* $a,b \in \mathbb{Z}$. *When* $d \equiv 1$ (mod 4) *the integers of* $\mathbb{Q}(\sqrt{d})$ *are the* $a+b\sqrt{d}$ *with* $a,b \in \mathbb{Z}$ *or* $a+1/2, b+1/2 \in \mathbb{Z}$.

Proof. If $a+b\sqrt{d} \in \mathbb{Q}(\sqrt{d})$ is an algebraic integer, and hence a solution of some equation $x^2 + Ax + B = 0$ with $A, B \in \mathbb{Z}$, then it follows from the quadratic formula that the other solution is $a - b\sqrt{d}$. Hence

$$x^2 + Ax + B = \left(x - (a+b\sqrt{d})\right)\left(x - (a-b\sqrt{d})\right).$$

Equating coefficients we get

$$A = -2a, \quad B = a^2 - db^2,$$

which shows that $2a$ and $a^2 - db^2$ are ordinary integers.

In particular, $a \in \mathbb{Z}$ or $a + 1/2 \in \mathbb{Z}$. In the first case ($2a$ even),

$$a \in \mathbb{Z} \Rightarrow a^2 \in \mathbb{Z}$$
$$\Rightarrow db^2 \in \mathbb{Z} \quad \text{since } a^2 - db^2 \in \mathbb{Z}$$
$$\Rightarrow b^2 \in \mathbb{Z} \quad \text{since no integer square } n^2 > 1 \text{ divides } d,$$
$$\text{and therefore } b^2 \neq m^2/n^2 \text{ with } n^2 > 1,$$
$$\Rightarrow b \in \mathbb{Z}.$$

The case $a + 1/2 \in \mathbb{Z}$ is when $2a$ is odd, so $(2a)^2 \equiv 1 \pmod 4$, and then

$$a^2 - db^2 \in \mathbb{Z} \Rightarrow (2a)^2 - d(2b)^2 \equiv 0 \pmod 4$$
$$\Rightarrow d(2b)^2 \equiv (2a)^2 \equiv 1 \pmod 4$$
$$\Rightarrow d \equiv 1 \pmod 4 \text{ and } (2b)^2 \equiv 1 \pmod 4$$
$$\text{since } (2b)^2 \equiv 3 \pmod 4 \text{ is impossible}$$
$$\Rightarrow d \equiv 1 \pmod 4 \text{ and } 2b \equiv 1 \pmod 2$$
$$\Rightarrow d \equiv 1 \pmod 4 \text{ and } b + 1/2 \in \mathbb{Z}.$$

Finally, to see that *every* number $a + b\sqrt{d}$ with $a + 1/2, b + 1/2 \in \mathbb{Z}$ is an integer of $\mathbb{Q}(\sqrt{d})$ when $d = 4m + 1$, it suffices to check the coefficients of the equation it satisfies: $x^2 - 2ax + (a^2 - db^2) = 0$. They are easily shown to be integers. $\qquad \square$

Exercises

10.4.1 Show that each element of $\mathbb{Q}(\sqrt{d})$ is an integer of $\mathbb{Q}(\sqrt{d})$ divided by an ordinary integer.

The second solution, α', of the monic quadratic equation with integer coefficients satisfied by a quadratic integer α is called the *conjugate* of α. This generalizes the notion of "complex conjugate" in $\mathbb{Z}[i]$ and the notion of "conjugate surd" from high school algebra.

10.4.2 If α is rational, what is α'?

10.4.3 Verify that conjugation is a *ring automorphism* of the integers of $\mathbb{Q}[\sqrt{d}]$, that is

- The map $\alpha \mapsto \alpha'$ is 1-to-1 and onto.
- $(\alpha + \beta)' = \alpha' + \beta'$ and $(\alpha\beta)' = \alpha'\beta'$.

10.5   Norm and units of quadratic fields

The *norm* on $\mathbb{Q}(\sqrt{d})$ is the function defined by

$$\text{norm}(a+b\sqrt{d}) = a^2 - db^2.$$

It follows that *an integer $a+b\sqrt{d}$ of $\mathbb{Q}(\sqrt{d})$ has (ordinary) integer norm*, because the proof of the last theorem of the previous section showed that $a^2 - db^2$ is an ordinary integer in that case. This norm includes the norms previously defined for $d = -1$, $d = -2$, and $d = n$, and like them it is *multiplicative*:

$$\text{norm}(x_1 x_2) = \text{norm}(x_1)\text{norm}(x_2) \quad \text{for any } x_1, x_2 \in \mathbb{Q}(\sqrt{d}).$$

This can be checked by setting $x_1 = a_1 + b_1\sqrt{d}$, $x_2 = a_2 + b_2\sqrt{d}$ and working out both sides. One finds the identity

$$(a_1 a_2 + db_1 b_2)^2 - d(a_1 b_2 + a_2 b_1)^2 = (a_1^2 - db_1^2)(a_2^2 - db_2^2),$$

which is Brahmagupta's identity of Section 5.4 when $d > 0$ and Diophantus' identity of Section 1.8 when $d = -1$. These properties of the norm imply that, *if x_1 divides x_2 in the integers of $\mathbb{Q}(\sqrt{d})$, then $\text{norm}(x_1)$ divides $\text{norm}(x_2)$ in the ordinary integers.*

As in any ring, the *units* among the integers of $\mathbb{Q}(\sqrt{d})$ are the elements that divide 1. It follows, by the previous remark, that the units of $\mathbb{Q}(\sqrt{d})$ are integers with norm ± 1. Conversely, integers of norm ± 1 are units because if $a + b\sqrt{d}$ is an integer (with $a, b \in \mathbb{Z}$ or $a + 1/2, b + 1/2 \in \mathbb{Z}$) with norm ± 1 then

$$\pm 1 = a^2 - db^2 = (a - b\sqrt{d})(a + b\sqrt{d}),$$

which shows that $a + b\sqrt{d}$ divides 1.

When $d > 1$ there are infinitely many units among the integers of $\mathbb{Q}(\sqrt{d})$, corresponding to the infinitely many solutions of the Pell equation $x^2 - dy^2 = 1$. For example, the solutions of $x^2 - 2y^2 = 1$ are pairs (x, y) for which $x + y\sqrt{2}$ is a unit of $\mathbb{Q}(\sqrt{2})$. We found these solutions in Section 5.2, giving the units $\pm(3 + 2\sqrt{2})^n$ for $n \in \mathbb{Z}$.

On the other hand, if $d < 0$ then there are only finitely many integers or half integers a, b with $a^2 - db^2 = 1$, and hence only finitely many units. In particular:

- the units of $\mathbb{Z}[i]$ are $\pm 1, \pm i$,

- the units of $\mathbb{Z}[\sqrt{-2}]$ are ± 1,

- the units of $\mathbb{Z}[\zeta_3]$, where $\zeta_3 = (-1 + \sqrt{-3})/2$, are $\pm 1, \pm \zeta_3, \pm \zeta_3^2$.

By the theorem in the previous section, $\mathbb{Z}[\zeta_3]$ is the ring of integers of the field $\mathbb{Q}(\sqrt{-3})$, and in fact it has the most units of any ring of integers of $\mathbb{Q}(\sqrt{d})$ with $d < 0$.

The fields $\mathbb{Q}(\sqrt{d})$ with $d < 0$ are called *imaginary quadratic fields* and their basic theory is particularly elegant. One advantage they have over real quadratic fields is that *when $d < 0$, $norm(a + b\sqrt{d}) = a^2 - db^2$ is simply the square of the distance of $a + b\sqrt{d}$ from 0 in the complex plane*. This makes certain properties geometrically obvious, such as the division property for $\mathbb{Z}[i]$ found in Section 6.4. Another example is the following theorem.

Units of imaginary quadratic fields. *The only units among the integers of imaginary quadratic fields are ± 1, $\pm i$, $\pm \zeta_3$, and $\pm \zeta_3^2$.*

Proof. Since units have norm 1, units of an imaginary quadratic field $\mathbb{Q}(\sqrt{d})$ lie at distance 1 from 0 in the complex plane. But we also know that the integers of $\mathbb{Q}(\sqrt{d})$ are of the form $a + b\sqrt{d}$ where $a, b \in \mathbb{Z}$ or $a + 1/2, b + 1/2 \in \mathbb{Z}$. If $|d| \geq 5$, all such integers except 0, ± 1 are at distance > 1 from 0. Thus the only units apart from ± 1 are those listed above, occurring in $\mathbb{Q}(i)$ and $\mathbb{Q}(\sqrt{-3})$. $\square$

The imaginary quadratic fields are better understood than the real ones. For example, Gauss (1801) found that the integers of $\mathbb{Q}(\sqrt{d})$ have unique prime factorization for $d = -1, -2, -3, -7, -11, -19, -43, -67, -163$, and in 1967 Baker and Stark showed that these are the *only* imaginary quadratic fields with unique prime factorization. The real quadratic fields with unique prime factorization are still not known, nor is it known whether there are infinitely many of them.

Exercises

An equivalent way to define the norm, which generalizes to arbitrary algebraic number fields of finite degree, is in terms of conjugates.

10.5.1 Show that $norm(\alpha) = \alpha \alpha'$.

10.5.2 If α is rational, what is $norm(\alpha)$?

10.5.3 Hence deduce the multiplicative property of the norm from the multiplicative property of conjugation.

10.6 Discussion

The rings $\mathbb{Z}$, $\mathbb{Q}$, $\mathbb{R}$, and $\mathbb{C}$ were known before the ring concept had a name. Rings began to proliferate in the mid-19th century when Kummer studied the cyclotomic rings $\mathbb{Z}[\zeta_n]$ and Dedekind sought a general theory of algebraic integers. Dedekind's first account of his theory appeared as a supplement to Dirichlet's *Vorlesungen über Zahlentheorie* (lectures on number theory) in 1871. At this time all the examples of interest to Dedekind were rings of algebraic numbers, so a ring (still unnamed) in 1871 could be defined as a subset of $\mathbb{C}$ closed under $+$, $-$, and $\times$.

The need for a definition by *axioms*, rather than closure properties, was gradually felt as other sets with $+$ and $\times$ operations came to be studied intensively. The congruence class rings $\mathbb{Z}/n\mathbb{Z}$ (defined in essentially the modern way by Dedekind (1857), with congruence classes as the objects being added and multiplied) were one class of examples. Another was the class of matrix rings, which were shown to include the quaternions by Cayley (1858) and many other structures by Peirce (1881).

Matrix rings, while generally noncommutative, also include many interesting commutative rings. We saw how $\mathbb{C}$ can be represented by 2×2 real matrices in Section 8.1. It is also possible to represent $\mathbb{Z}[\alpha]$, for any algebraic integer α of degree n, by $n \times n$ integer matrices, and $\mathbb{Q}(\alpha)$ by $n \times n$ rational matrices. Briefly, the idea is this.

If the monic equation satisfied by α is

$$\alpha^n + a_{n-1}\alpha^{n-1} + \cdots + a_1\alpha + a_0 = 0 \quad \text{with} \quad a_0,\ldots,a_{n-1} \in \mathbb{Z}$$

then $\alpha^n = -a_{n-1}\alpha^{n-1} - \cdots - a_1\alpha - a_0$ and hence all powers $\alpha^n, \alpha^{n+1} \ldots$ are rational linear combinations of $1, \alpha, \alpha^2, \ldots, \alpha^{n-1}$. This allows $\mathbb{Q}(\alpha)$ to be viewed as a *vector space over* $\mathbb{Q}$ with basis $\{1, \alpha, \alpha^2, \ldots, \alpha^{n-1}\}$. Multiplication by α induces a linear map of this vector space with matrix

$$M_\alpha = \begin{pmatrix} 0 & 0 & \cdots & 0 & -a_0 \\ 1 & 0 & \cdots & 0 & -a_1 \\ 0 & 1 & \cdots & 0 & -a_2 \\ \vdots & & \ddots & & \vdots \\ 0 & 0 & \cdots & 1 & -a_{n-1} \end{pmatrix},$$

because right multiplication of the row vector $(1 \ \alpha \ \alpha^2 \ \cdots \ \alpha^{n-2} \ \alpha^{n-1})$ by M_α yields its multiple by α:

$$(\alpha \ \alpha^2 \ \alpha^3 \ \cdots \ -a_{n-1}\alpha^{n-1} - \cdots - a_1\alpha - a_0).$$

It follows that matrix polynomials in M_α, with rational coefficients, behave the same as the corresponding polynomials in α, that is, the same as elements of $\mathbb{Q}(\alpha)$.

All of these examples were finally unified under the abstract ring concept, defined axiomatically by Fraenkel (1914) and developed by Emmy Noether and her students in the 1920s. Noether always said "Es steht schon bei Dedekind" ("It's already in Dedekind"), and she urged her students to read all of Dedekind's works on algebraic number theory. These included three different versions of his last supplement to Dirichlet's *Vorlesungen*, in 1871, 1879, and 1894, and a separately published work that is now available in English translation, Dedekind (1877). The latter is probably the easiest introduction to Dedekind's work for the modern reader.

Among other things, Dedekind (1877) generalizes the concept of norm to an arbitrary algebraic number field $\mathbb{Q}(\alpha)$. He defines the conjugates $\alpha', \alpha'', \ldots$ of α to be the other solutions of the minimal degree monic equation satisfied by α and defines norm(α) as the product $\alpha\alpha'\alpha'' \cdots$. It can then be shown that norm(α) is an ordinary integer when α is an integer of $\mathbb{Q}(\alpha)$, and that norm($\alpha\beta$) = norm(α)norm(β). The proofs are not hard but the latter requires concepts from the companion volume to this one, *Elements of Algebra* (field isomorphisms). It can be said that algebraic number theory is where the concepts from the theory of algebraic equations (groups and fields) begin to interact with concepts from the theory of Diophantine equations (rings).

11

Ideals

PREVIEW

This chapter pursues the idea that a number is known by the set of its multiples, so an "ideal number" is known by a set that *behaves like* a set of multiples. Such a set I in a ring R is called an *ideal*, and it is defined by closure under sums $(a, b \in I \Rightarrow a + b \in I)$ and under multiplication by all elements of the ring $(a \in I, r \in R \Rightarrow ar \in I)$.

The set $(a) = \{ar : r \in R\}$ of all multiples of any $a \in R$ is an ideal, called the *principal ideal generated by a*. Any nonprincipal ideal I is therefore *not* the set of multiples of any actual member of R—it represents an "ideal member" of R.

In $\mathbb{Z}$, every ideal is principal, and the properties of ideals reflect known properties of integers. In particular:

a divides $b \Leftrightarrow (a)$ contains (b)

$(\gcd(a, b)) = \{am + bn : m, n \in \mathbb{Z}\}$, the ideal generated by a and b

p is prime $\Leftrightarrow (p)$ is *maximal*.

In rings where unique prime factorization fails, such as $\mathbb{Z}[\sqrt{-5}]$, nonprincipal ideals exist. We find such an ideal as the "gcd ideal" of 2 and $1 + \sqrt{-5}$, $\{2m + (1 + \sqrt{-5})n : m, n \in \mathbb{Z}\}$, and confirm that this ideal is nonprincipal by looking at its shape in the plane.

The ideal $\{2m + (1 + \sqrt{-5})n : m, n \in \mathbb{Z}\}$ "divides" the principal ideal (2) because it contains it, and it is "prime" because it is maximal. But if an ideal I "divides" an ideal J there should be a "product" IK of ideals I and K equal to J.

We hope that (2) splits into such a product, because this may rectify the failure of unique prime factorization in $\mathbb{Z}[\sqrt{-5}]$ exhibited by $2 \times 3 = (1 + \sqrt{-5})(1 - \sqrt{-5})$.

196

Thus the final step is to define the product of ideals. It then turns out, as hoped, that the two distinct products equal to 6 in $\mathbb{Z}[\sqrt{-5}]$, 2×3 and $(1 + \sqrt{-5})(1 - \sqrt{-5})$, split into the *same* product of prime ideals.

11.1   Ideals and the gcd

In Section 7.4 we had our first brush with failure of unique prime factorization when we found that 4 has two distinct prime factorizations in $\mathbb{Z}[\sqrt{-3}]$:

$$4 = 2 \times 2 = (1 + \sqrt{-3})(1 - \sqrt{-3}).$$

This problem was fixed by enlarging $\mathbb{Z}[\sqrt{-3}]$ to $\mathbb{Z}[\frac{-1+\sqrt{-3}}{2}]$, where the factorizations 2×2 and $(1 + \sqrt{-3})(1 - \sqrt{-3})$ are actually the *same*, up to unit factors. This is because $\mathbb{Z}[\frac{-1+\sqrt{-3}}{2}]$ contains the units $\frac{1+\sqrt{-3}}{2}$ and $\frac{1-\sqrt{-3}}{2}$ whose product is 1, and therefore

$$2 \times 2 = 2 \left(\frac{1 + \sqrt{-3}}{2} \right) 2 \left(\frac{1 - \sqrt{-3}}{2} \right) = (1 + \sqrt{-3})(1 - \sqrt{-3}).$$

However, this is in some sense a lucky escape, and a more serious problem occurs in $\mathbb{Z}[\sqrt{-5}]$, where 6 has two different factorizations:

$$6 = 2 \times 3 = (1 + \sqrt{-5})(1 - \sqrt{-5}).$$

Using the norm $a^2 + 5b^2$ of any $a + b\sqrt{-5} \in \mathbb{Z}[\sqrt{-5}]$, it can be checked that none of these factors are products of elements of smaller norm. Nor are the units ± 1 of $\mathbb{Z}[\sqrt{-5}]$ able to account for the difference between the factorizations. Thus 6 *has two distinct prime factorizations in* $\mathbb{Z}[\sqrt{-5}]$. And we cannot get around this problem by a simple enlargement of the ring, like that of $\mathbb{Z}[\sqrt{-3}]$ to $\mathbb{Z}[\frac{-1+\sqrt{-3}}{2}]$, because $\mathbb{Z}[\sqrt{-5}]$ already contains all the integers of $\mathbb{Q}(\sqrt{-5})$.

In such situations, Kummer and Dedekind were able to restore unique prime factorization by extending the concepts of product and divisibility to what Kummer called "ideal numbers" and what Dedekind called *ideals*. We find our first "ideal number" by searching for the gcd of 2 and $1 + \sqrt{-5}$ in $\mathbb{Z}[\sqrt{-5}]$, based on a new approach to the gcd in $\mathbb{Z}$.

The gcd in $\mathbb{Z}$ revisited

The basic idea of Kummer and Dedekind is that *a number is known by the set of its multiples*. We illustrate this idea, and its application to the gcd, by the example of $\gcd(4,6)$. Figure 11.1 shows the set of multiples of 4, which we denote by (4), as black dots among the integers:

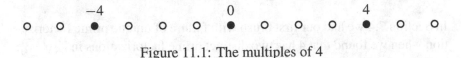

Figure 11.1: The multiples of 4

Similarly, Figure 11.2 shows the set (6) of multiples of 6:

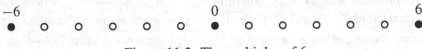

Figure 11.2: The multiples of 6

Finally, Figure 11.3 shows all sums of members of (4) and members of (6). We denote the set of these sums by $(4)+(6)$:

Figure 11.3: The sums of multiples of 4 and multiples of 6

It is clear that $(4)+(6)=(2)$ and that $2=\gcd(4,6)$ so *the multiples of* $\gcd(4,6)$ *are obtained by adding all multiples of 4 to all multiples of 6.* More generally, we let (k) denote the set $\{kn : n \in \mathbb{Z}\}$ of multiples of k for any $k \in \mathbb{Z}$. Then we have: *the set* $(a)+(b)=\{am+bn : m,n \in \mathbb{Z}\}$ *equals the set* $(\gcd(a,b))$ *of multiples of* $\gcd(a,b)$.

We prove this theorem in the next section. It gives an abstract alternative to the Euclidean algorithm for finding the gcd, with the advantage of being applicable to any ring R. This is done by replacing the sets (k) above by the following more general concept:

Definition. An *ideal* in a ring R is a subset I of R such that

- $a \in I$ and $b \in I \Rightarrow a+b \in I$,

- $a \in I$ and $r \in R \Rightarrow ar \in I$.

In other words, I is closed under addition and closed under multiplication by elements of R. It follows that I is also closed under subtraction, because $b \in I \Rightarrow -b \in I$ (multiplication by $-1 \in R$) and $a - b = a + (-b)$.

It is often the case that if I and J are ideals, then so is

$$I + J = \{i + j : i \in I \text{ and } j \in J\}.$$

The latter is what we call $\gcd(I, J)$, and we use it in Section 11.5 to find the gcd of 2 and $1 + \sqrt{-5}$ in $\mathbb{Z}[\sqrt{-5}]$. But first we investigate the concept of ideal in the more familiar ring $\mathbb{Z}$.

Exercises

The theorem that $(a) + (b) = (\gcd(a, b))$ in $\mathbb{Z}$ can be proved directly, and it is worth doing so in advance of the proof using ideal theory in the next section.

11.1.1 Show that all members of $(a) + (b) = \{am + bn : m, n \in \mathbb{Z}\}$ are multiples of $\gcd(a, b)$.

11.1.2 Show that $\{am + bn : m, n \in \mathbb{Z}\}$ is closed under difference, and deduce that all its members are multiples of the smallest positive member.

11.1.3 Deduce that $c = \gcd(a, b)$, and hence that $(a) + (b) = (\gcd(a, b))$.

11.2 Ideals and divisibility in $\mathbb{Z}$

The simplest examples of ideals occur in $\mathbb{Z}$. For instance $(2) = \{2n : n \in \mathbb{Z}\}$ is an ideal, as is $(6) = \{6n : n \in \mathbb{Z}\}$. In fact, for any $a \in \mathbb{Z}$, the set

$$(a) = \{an : n \in \mathbb{Z}\}$$

is an ideal called the *principal ideal generated by a*. Divisibility in $\mathbb{Z}$ corresponds to *containment* of principal ideals. For example

$$2 \text{ divides } 6, \quad (2) \text{ contains } (6),$$

and in general

$$a \text{ divides } b \quad \Leftrightarrow \quad (a) \text{ contains } (b).$$

Dedekind's idea was to *define* "divisibility" of ideals in a ring R by saying that

$$I \text{ "divides" } J \quad \Leftrightarrow \quad I \text{ contains } J.$$

(We put this notion of division in quotes for now, because it remains to be seen whether it is consistent with the usual notion of divisibility: I divides $J \Leftrightarrow J = IK$ for some ideal K. The latter notion comes into contention when we define the product of ideals in Section 11.7.)

Dedekind's definition includes divisibility of elements $s, t \in R$, since in any ring we can define the principal ideals

$$(s) = \{sr : r \in R\}, \quad (t) = \{tr : r \in R\}$$

and show that

$$s \text{ divides } t \quad \Leftrightarrow \quad (s) \text{ contains } (t).$$

But "divisibility" of ideals generally *extends* the concept of divisibility of elements, because not every ideal is principal.

In particular, we shall see that there are nonprincipal ideals in $\mathbb{Z}[\sqrt{-5}]$, and they include "ideal primes" that restore unique prime factorization. Before doing so, however, it is helpful to look more closely at $\mathbb{Z}$ from the viewpoint of ideals. This allows us to see how the basic theory of ideals elegantly includes the traditional theory of divisibility, common divisors, and primes.

Ideal theory in $\mathbb{Z}$

The basic theory of divisibility in $\mathbb{Z}$ consists of three theorems about ideals. The first is a counterpart of the division property.

Principal ideal property of $\mathbb{Z}$. *All ideals in $\mathbb{Z}$ are principal.*

Proof. Suppose I is an ideal of $\mathbb{Z}$ other than (0). Then I has a least positive member, a say. Since I is closed under multiplication by members of $\mathbb{Z}$, I includes all members of

$$(a) = \{an : n \in \mathbb{Z}\}.$$

But these are the *only* members of I, because if b is not a multiple of a then

$$b - (\text{greatest multiple of } a \text{ less than } b)$$

is a positive member of I less than a, contrary to assumption. $\square$

The second theorem yields the gcd without the Euclidean algorithm, generalizing the example of the previous section.

The gcd ideal. *The set* $(a) + (b) = \{am + bn : m, n \in \mathbb{Z}\}$ *is* $(\gcd(a, b))$.

Proof. Since $\gcd(a, b)$ divides a and b, it divides each number $am + bn$. Thus $\{am + bn : m, n \in \mathbb{Z}\}$ includes only multiples of $\gcd(a, b)$.

Now $\{am + bn : m, n \in \mathbb{Z}\}$ is clearly an ideal, hence by the previous theorem it consists of the multiples of its least positive member c. Since a and b are numbers of the form $am + bn$, they too are multiples of c, hence c is a common divisor of a and b. But we already know that c is a multiple of $\gcd(a, b)$, hence $c = \gcd(a, b)$. Thus the ideal $\{am + bn : m, n \in \mathbb{Z}\}$ of multiples of c includes exactly the multiples of $\gcd(a, b)$. □

Finally, we can express the prime divisor property in terms of ideals. The proof is close to the one originally given for the prime divisor property (Section 2.4). Since "divides" means "contains" for ideals, the only ideals containing an ideal (p) for prime p are (p) itself and $(1) = \mathbb{Z}$.

Prime ideal property. *If p is prime and the ideal (p) contains (ab), then (p) contains (a) or (p) contains (b).*

Proof. Suppose $(a) \nsubseteq (p)$, so we have to prove $(b) \subseteq (p)$.

Since the ideal $\{am + pn : m, n \in \mathbb{Z}\}$ contains both (p) and (a), and $(a) \nsubseteq (p)$, $\{am + pn : m, n \in \mathbb{Z}\}$ can only equal (1).

This means $1 = am + pn$ for some $m, n \in \mathbb{Z}$ and

$$1 = am + pn \Rightarrow b = abm + pbn \quad \text{multiplying both sides by } b$$
$$\Rightarrow b \in (p) \quad \text{since } ab \in (p) \text{ by hypothesis and } p \in (p)$$
$$\Rightarrow (b) \subseteq (p) \quad \text{as required} \qquad \square$$

As we know, the prime divisor property is the essence of unique prime factorization. However, we cannot discuss factorization of general ideals yet, because we have not defined the product of ideals. Likewise, we cannot yet define the general concept of "prime" ideal, though the prime ideal property for $\mathbb{Z}$ suggests how it should be done as soon as we have defined the product.

Exercises

Just as $\gcd(a, b)$ results from a simple operation on (a) and (b), so does $\text{lcm}(a, b)$. Moreover, the operation makes sense on any ideals, and hence gives a general definition of lcm in any ring.

11.2.1 Show that $(\text{lcm}(a, b)) = (a) \cap (b)$.

11.2.2 For any ideals I and J in a ring R, prove that $I \cap J$ is an ideal.

11.3 Principal ideal domains

$\mathbb{Z}$ is an example of a *principal ideal domain*, a ring R in which every ideal is of the form $(a) = \{ar : r \in R\}$.

Other examples are rings with a Euclidean algorithm, such as $\mathbb{Z}[i]$, $\mathbb{Z}[\sqrt{-2}]$, and $\mathbb{Z}[\zeta_3]$. As we know, a ring R has a Euclidean algorithm if R has the division property, which we can formalize as follows.

Definition. R is a *Euclidean ring* if there is a nonnegative integer-valued function $|r|$ on R such that $|r| = 0$ only if $r = 0$, and for any $a, b \in R$ with $|b| > 0$ there are $q, r \in R$ such that $a = qb + r$ with $0 \leq |r| < |b|$.

For the rings mentioned above, the function $|r|$ is the absolute value function. For certain examples where the norm can be negative, such as $\mathbb{Z}[\sqrt{2}]$, the square root of the absolute value of the norm will serve as $|r|$. (Exercise 9.1.6 is then a proof that $\mathbb{Z}[\sqrt{2}]$ is a Euclidean ring.)

The theorems about $\mathbb{Z}$ in the previous section generalize to the following theorems about principal ideal domains. Since the proofs are similar, we abbreviate them slightly here.

Principal ideal property of Euclidean rings. *Any Euclidean ring is a principal ideal domain.*

Proof. If $I \neq (0)$ is an ideal of R let $b \in I$ be of minimal norm > 0. Since I is an ideal it includes all multiples of b by elements of R.

Conversely, if $a \in I$ is not a multiple of b, then we have $a = qb + r$ with $0 < |r| < |b|$, and $r = a - qb \in I$, contrary to the minimality of b.

Thus $I = (b)$.   □

Prime divisor property for principal ideal domains. *If p is a prime in a principal ideal domain and p divides ab, then p divides a or p divides b.*

Proof. Suppose p is a prime that divides ab but not a, so we have to prove that p divides b.

R a principal ideal ring $\Rightarrow$ ideal $\{ar + ps : r, s \in R\} = (t)$ for some $t \in R$

$\qquad\qquad \Rightarrow (t) \supseteq (a)$ and $(t) \supseteq (p)$

$\qquad\qquad \Rightarrow t$ divides a and p

$\qquad\qquad \Rightarrow t = 1$ since p is prime

$\qquad\qquad \Rightarrow 1 = ar + ps$ for some $r, s \in R$

$\qquad\qquad \Rightarrow b = abr + pbs$

$\qquad\qquad \Rightarrow p$ divides b.   □

These theorems give a uniform explanation why the prime divisor property (and hence unique prime factorization) holds in the rings $\mathbb{Z}$, $\mathbb{Z}[i]$, $\mathbb{Z}[\sqrt{-2}]$, and $\mathbb{Z}[\zeta_3]$—they are all Euclidean, and hence they are principal ideal domains.

The principal ideal property of $\mathbb{Z}[i]$ has an interesting geometric interpretation. As we observed in Section 6.4, the principal ideal (β) of all multiples of $\beta \neq 0$ in $\mathbb{Z}[i]$ is a lattice of the same shape as $\mathbb{Z}[i]$ (and magnified by $|\beta|$). Therefore, since all ideals of $\mathbb{Z}[i]$ are principal, it follows that *any ideal $\neq (0)$ in $\mathbb{Z}[i]$ has the same shape as $\mathbb{Z}[i]$.* The same is true of $\mathbb{Z}[\sqrt{-2}]$ and $\mathbb{Z}[\zeta_3]$, as we observed in Sections 7.2 and 7.4. (We say that two sets in the plane have the "same shape" if one can be mapped onto the other by a function that multiplies all distances by a constant.)

Conversely, nonprincipal ideals exist only when unique prime factorization fails, and therefore *differently shaped ideals exist only when unique prime factorization fails.* We shall see that they actually occur in $\mathbb{Z}[\sqrt{-3}]$ and $\mathbb{Z}[\sqrt{-5}]$.

Exercises

It happens that the only imaginary quadratic fields $\mathbb{Q}(\sqrt{d})$ whose integers form Euclidean rings are those with $d = -1, -2, -3, -7, -11$. The only two we have not studied already are $\mathbb{Q}(\sqrt{-7})$ and $\mathbb{Q}(\sqrt{-11})$, the integer rings of which are $\mathbb{Z}[\frac{1+\sqrt{-7}}{2}]$ and $\mathbb{Z}[\frac{1+\sqrt{-11}}{2}]$ respectively, because $d \equiv 1 \pmod 4$ in these cases. We let $\omega = \frac{1+\sqrt{-7}}{2}$ or $\frac{1+\sqrt{-11}}{2}$ and consider the points of $\mathbb{Z}[\omega]$ in the plane $\mathbb{C}$.

As usual (see Chapter 7), the division property is implied by the statement that any point of the plane is at distance < 1 from the nearest point of $\mathbb{Z}[\omega]$. We prove this with the help of Figure 11.4, which shows 0 and its 6 neighbors ± 1, $\pm \omega$, $\pm(\omega - 1)$. The hexagonal region around 0 is bounded by the perpendicular bisectors of the lines from 0 to its neighbors, and hence the points inside it are those that are nearer to 0 than to any neighbor. Since any point of $\mathbb{Z}[\omega]$ looks like 0, it suffices to prove that the point on the hexagon farthest from 0, namely ai, is at distance < 1.

11.3.1 The point ai is equidistant from 0 and ω (why?). Deduce that

$$|ai| = |ai - \omega| = \left| -\frac{1}{2} + \frac{2a - \sqrt{|d|}}{2} i \right|.$$

11.3.2 Deduce from Exercise 11.3.1 that $a = \frac{1+|d|}{4\sqrt{|d|}}$.

11.3.3 Deduce from Exercise 11.3.2 that $a < 1$ for $d = -7, -11$, and hence that $\mathbb{Z}[\frac{1+\sqrt{d}}{2}]$ has the division property in those cases.

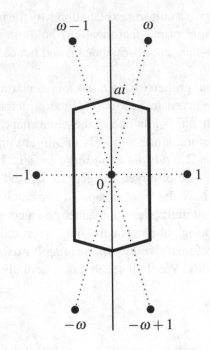

Figure 11.4: The region of points nearer to 0 than its neighbors

Thus $\mathbb{Z}[\frac{1+\sqrt{-7}}{2}]$ and $\mathbb{Z}[\frac{1+\sqrt{-11}}{2}]$ have unique prime factorization, which we can apply to solve the equations $y^3 = x^2 + 7$ and $y^3 = x^2 + 11$ as we solved $y^3 = x^2 + 1, x^2 + 2$, and $x^2 + 4$ in Chapter 7 and its exercises. The case of $y^3 = x^2 + 11$ is particularly interesting, because one of its solutions is large enough not to be obvious.

11.3.4 If $y^3 = x^2 + 11 = (x + \sqrt{-11})(x - \sqrt{-11})$, use unique prime factorization in $\mathbb{Z}[\frac{1+\sqrt{-11}}{2}]$ to show that

$$x + \sqrt{-11} = (a^3 - 33ab^2) + (3a^2b - 11b^3)\sqrt{-11}$$

where a, b are both integers or else both half-integers that are not integers.

11.3.5 Find one integer solution of $y^3 = x^2 + 11$ by finding an integer solution of the equation $b(3a^2 - 11b^2) = 1$.

11.3.6 Find another integer solution of $y^3 = x^2 + 11$ by finding a half-integer solution of $b(3a^2 - 11b^2) = 1$, and show that it is the only other integer solution x, y (up to the sign of x).

11.4 A nonprincipal ideal of $\mathbb{Z}[\sqrt{-3}]$

We might attempt to understand the nonunique prime factorization of 4 in $\mathbb{Z}[\sqrt{-3}]$,

$$4 = 2 \times 2 = (1 + \sqrt{-3})(1 - \sqrt{-3}),$$

by looking for an "ideal" common divisor of 2 and $1 + \sqrt{-3}$ to split the factors. Using the idea of Section 11.1, we form the ideal $(2) + (1 + \sqrt{-3})$ by adding all multiples of 2 to all multiples of $1 + \sqrt{-3}$.

An arbitrary element of $(2) + (1 + \sqrt{-3})$ is

$$2(a + b\sqrt{-3}) + (1 + \sqrt{-3})(c + d\sqrt{-3}) \quad \text{for some} \quad a, b, c, d \in \mathbb{Z}$$
$$= 2(a - b - 2d) + (1 + \sqrt{-3})(2b + c + d)$$
$$= 2m + (1 + \sqrt{-3})n \quad \text{for some} \quad m, n \in \mathbb{Z}.$$

Conversely, $2m + (1 + \sqrt{-3})n \in (2) + (1 + \sqrt{-3})$ for any $m, n \in \mathbb{Z}$, and therefore

$$(2) + (1 + \sqrt{-3}) = \{2m + (1 + \sqrt{-3})n : m, n \in \mathbb{Z}\}.$$

Figure 11.5 shows the elements of this ideal, marked in black, on a picture of $\mathbb{Z}[\sqrt{-3}]$. It is clear that $(2) + (1 + \sqrt{-3})$ consists of equilateral triangles, and hence it is not of the same shape as $\mathbb{Z}[\sqrt{-3}]$. (For example, no point of the ideal has neighbors in perpendicular directions.)

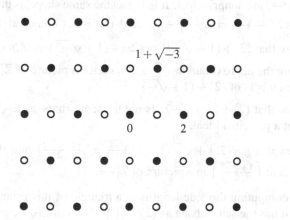

Figure 11.5: The ideal $(2) + (1 + \sqrt{-3})$ in $\mathbb{Z}[\sqrt{-3}]$

Thus $(2) + (1 + \sqrt{-3})$ *is a nonprincipal ideal*: it does *not* consist of the multiples of any member of $\mathbb{Z}[\sqrt{-3}]$. However, we can dream that $(2) + (1 + \sqrt{-3})$ consists of multiples of an "ideal number"—something outside $\mathbb{Z}[\sqrt{-3}]$—and this dream is easily realized.

The shape of $(2) + (1 + \sqrt{-3})$ is exactly the same as the shape of $\mathbb{Z}[\frac{1+\sqrt{-3}}{2}]$, the set of multiples of $\frac{1+\sqrt{-3}}{2}$ (Figure 11.6), so $\frac{1+\sqrt{-3}}{2}$ is the desired "ideal number". In $\mathbb{Z}[\frac{1+\sqrt{-3}}{2}]$, $\frac{1+\sqrt{-3}}{2}$ divides both 2 and $1 + \sqrt{-3}$, and its norm is 1, hence $\frac{1+\sqrt{-3}}{2}$ is the gcd of 2 and $1 + \sqrt{-3}$.

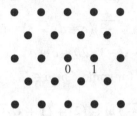

Figure 11.6: The principal ideal $\left(\frac{1+\sqrt{-3}}{2} \right) = \mathbb{Z}\left[\frac{1+\sqrt{-3}}{2} \right]$

Thus the "ideal number" whose "multiples" make up the nonprincipal ideal really exists in this case, but outside the original ring.

Exercises

An interesting variation of this phenomenon occurs in $\mathbb{Z}[\sqrt{-7}]$, where the ideal $(2) + (1 + \sqrt{-7})$ is nonprincipal. It is in fact the same shape as the principal ideal of multiples of $\frac{1+\sqrt{-7}}{2}$ in $\mathbb{Z}[\frac{1+\sqrt{-7}}{2}]$, though this is not immediately obvious.

11.4.1 Show that $(2) + (1 + \sqrt{-7}) = \{2m + (1 + \sqrt{-7})m : m, n \in \mathbb{Z}\}$.

11.4.2 Using the approximation $\sqrt{7} \simeq 2.6$, sketch a picture of $\mathbb{Z}[\sqrt{-7}]$ and mark the members of $(2) + (1 + \sqrt{-7})$.

11.4.3 Show that $(2) + (1 + \sqrt{-7})$ is not the same shape as $\mathbb{Z}[\sqrt{-7}]$, and hence is not a principal ideal.

11.4.4 Show that $\gcd(2, 1 + \sqrt{-7}) = \frac{1+\sqrt{-7}}{2}$ in $\mathbb{Z}[\frac{1+\sqrt{-7}}{2}]$, and sketch the principal ideal $\left(\frac{1+\sqrt{-7}}{2} \right)$ in a picture of $\mathbb{Z}[\frac{1+\sqrt{-7}}{2}]$.

11.4.5 By computing the side lengths in a triangle of this principal ideal, show that it has the same shape as $(2) + (1 + \sqrt{-7})$ in $\mathbb{Z}[\sqrt{-7}]$.

11.4.6 Show that 8 has distinct prime factorizations in $\mathbb{Z}[\sqrt{-7}]$, but that each of these factorizations splits into the same prime factorization in $\mathbb{Z}[\frac{1+\sqrt{-7}}{2}]$.

11.5 A nonprincipal ideal of $\mathbb{Z}[\sqrt{-5}]$

Like $\mathbb{Z}[\sqrt{-3}]$, $\mathbb{Z}[\sqrt{-5}]$ fails to have the prime divisor property, as is shown by the two factorizations

$$2 \times 3 = (1 + \sqrt{-5})(1 - \sqrt{-5}).$$

The numbers 2, 3, $1 + \sqrt{-5}$, $1 - \sqrt{-5}$ have norms 4, 9, 6, 6 respectively, whose divisors 2 and 3 are *not* the norm $a^2 + 5b^2$ of any member $a + b\sqrt{-5}$ of $\mathbb{Z}[\sqrt{-5}]$. Hence none of 2, 3, $1 + \sqrt{-5}$, $1 - \sqrt{-5}$ is a product of numbers of smaller norm, and so they are primes of $\mathbb{Z}[\sqrt{-5}]$.

To understand this nonunique prime factorization we first construct the "ideal" gcd of 2 and $1 + \sqrt{-5}$: the set $(2) + (1 + \sqrt{-5})$ of sums of multiples of 2 and multiples of $1 + \sqrt{-5}$. A similar calculation to the one in the previous section shows that any member

$$2(a + b\sqrt{-5}) + (1 + \sqrt{-5})(c + d\sqrt{-5}) \in (2) + (1 + \sqrt{-5})$$

is actually of the form $2m + (1 + \sqrt{-5})n$ for some $m, n \in \mathbb{Z}$ (exercise). Conversely, any such number $2n + (1 + \sqrt{-5})n$ is in $(2) + (1 + \sqrt{-5})$, so

$$(2) + (1 + \sqrt{-5}) = \{2m + (1 + \sqrt{-5})n : m, n \in \mathbb{Z}\}.$$

Figure 11.7 shows this ideal as black dots in a picture of $\mathbb{Z}[\sqrt{-5}]$. It is clear from the picture that no black dot has neighbors in perpendicular directions, so this ideal is not of the same shape as $\mathbb{Z}[\sqrt{-5}]$. In particular, it is not a principal ideal (β), since (β) is simply $\mathbb{Z}[\sqrt{-5}]$ multiplied by β, which multiplies all distances by $|\beta|$ and hence produces a set of the same shape, as observed in Section 8.1. Thus $(2) + (1 + \sqrt{-5})$ is nonprincipal.

We would like to view the members of $(2) + (1 + \sqrt{-5})$ as multiples of some "ideal number" outside $\mathbb{Z}[\sqrt{-5}]$. However, $\mathbb{Z}[\sqrt{-5}]$ already contains all the integers of $\mathbb{Z}[\sqrt{-5}]$, so it is not clear where this "ideal number" can be found. Instead, we follow Dedekind (1871) and *do without the "ideal number", working directly with the ideal.*

As we have seen, a principal ideal (β) behaves the same as the number β in the sense that

$$(\beta) \text{ contains } (\gamma) \Leftrightarrow \beta \text{ divides } \gamma.$$

In this sense, the nonprincipal ideal $(2) + (1 + \sqrt{-5})$ behaves like a number dividing both 2 and $1 + \sqrt{-5}$, because

$$(2) + (1 + \sqrt{-5}) \text{ contains } (2), \quad (2) + (1 + \sqrt{-5}) \text{ contains } (1 + \sqrt{-5}).$$

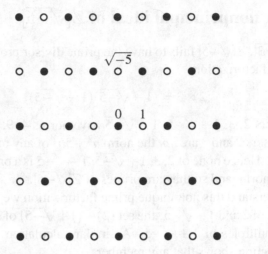

Figure 11.7: The nonprincipal ideal $(2)+(1+\sqrt{-5})$ of $\mathbb{Z}[\sqrt{-5}]$

And in fact we have seen that $(2)+(1+\sqrt{-5})$ deserves to be the *"greatest common divisor"* of 2 and $1+\sqrt{-5}$ in $\mathbb{Z}[\sqrt{-5}]$, since the analogous ideal in $\mathbb{Z}$, $(a)+(b)=\{am+bn:m,n\in\mathbb{Z}\}$, *is* $(\gcd(a,b))$, as we saw in Section 11.1.

Not only that. $(2)+(1+\sqrt{-5})$ deserves to be called a *prime*. We noted in Section 11.2 that a prime *principal* ideal (p) is maximal in the sense that (1) and (p) are the only ideals containing it. Likewise, the only ideals containing $(2)+(1+\sqrt{-5})$ are (1) and $(2)+(1+\sqrt{-5})$ itself. This is because

$$(2)+(1+\sqrt{-5})=\{2m+(1+\sqrt{-5})n:m,n\in\mathbb{Z}\}$$

consists of the numbers of the form even $+(1+\sqrt{-5})n$. Hence any member of $\mathbb{Z}[\sqrt{-5}]$ not in $(2,1+\sqrt{-5})$ is of the form odd $+(1+\sqrt{-5})n$. But an ideal including all numbers even $+(1+\sqrt{-5})n$ and *at least one* number odd $+(1+\sqrt{-5})n'$ obviously includes 1, hence all of $\mathbb{Z}[\sqrt{-5}]$ by closure under multiplication by members of $\mathbb{Z}[\sqrt{-5}]$. Thus $(2)+(1+\sqrt{-5})$ is a maximal ideal and we shall see, once we have defined the product of ideals, that maximal ideals are prime.

We also need to define the product of ideals to confirm that the ideal $(2)+(1+\sqrt{-5})$ divides (2) in the usual sense of ring theory, namely $(2)=((2)+(1+\sqrt{-5}))\times I$ for some ideal I. This will be done in Section 11.7, and we find the mystery factor I in Section 11.8.

Exercises

11.5.1 Check that $2(a+b\sqrt{-5})+(1+\sqrt{-5})(c+d\sqrt{-5})$, when $a,b,c,d \in \mathbb{Z}$, is of the form $2m+(1+\sqrt{-5})n$ for some $m,n \in \mathbb{Z}$.

Also important in reconciling the two prime factorizations of 6 in $\mathbb{Z}[\sqrt{-5}]$ is the gcd ideal of 3 and $1+\sqrt{-5}$, the general member of which is of the form

$$3(a+b\sqrt{-5})+(1+\sqrt{-5})(c+d\sqrt{-5}), \quad \text{where} \quad a,b,c,d \in \mathbb{Z}.$$

11.5.2 Check that $3(a+b\sqrt{-5})+(1+\sqrt{-5})(c+d\sqrt{-5})$, $a,b,c,d \in \mathbb{Z}$, is of the form $3m+(1+\sqrt{-5})n$ for some $m,n \in \mathbb{Z}$.

11.5.3 Deduce from Exercise 11.5.2 that

$$(3)+(1+\sqrt{-5}) = \{3m+(1+\sqrt{-5})n : m,n \in \mathbb{Z}\}.$$

11.6 Ideals of imaginary quadratic fields as lattices

Although ideals in rings of quadratic integers are not always principal, we can nonetheless prove an analogue of the theorem that $\mathbb{Z}$ is a principal ideal domain. Ideals I of $\mathbb{Z}$ each have *one generator*, in the sense that each consists of the integer multiples of some integer a. $\mathbb{Z}$ itself is the ideal with generator 1. The analogous theorem (with a similar proof) for ideals I in the integers of $\mathbb{Q}(\sqrt{d})$ says that they have *two* generators, like the integers of $\mathbb{Q}(\sqrt{d})$ themselves.

The description of the integers of $\mathbb{Q}(\sqrt{d})$ in Section 10.4 shows that they comprise either $\mathbb{Z}[\frac{1+\sqrt{d}}{2}]$ or $\mathbb{Z}[\sqrt{d}]$. In either case, they form a subgroup L of $\mathbb{C}$ with *two generators*: 1 and $\frac{1+\sqrt{d}}{2}$ for $\mathbb{Z}[\frac{1+\sqrt{d}}{2}]$, and 1 and $\sqrt{d}$ for $\mathbb{Z}[\sqrt{d}]$. When $d < 0$ the generators of L can be described geometrically as two nonzero members nearest to 0 but not on the same line through 0, and the group they generate is called a *lattice*.

In general, a lattice L in $\mathbb{C}$ is a set $\{\alpha m + \beta n : m,n \in \mathbb{Z}\}$ where α and β are nonzero complex numbers not on the same line through 0. The pair α, β of generators is called an *integral basis* for L. The elements of L lie at the intersections of two families of parallel lines forming a "lattice" in the ordinary sense of the word (Figure 11.8). An important property of the lattice of integers in $\mathbb{Z}(\sqrt{d})$ is that any subgroup of it, and hence any ideal, is also a lattice.

Lattice property of ideals. *When $d < 0$, any nonzero ideal in the integers of $\mathbb{Q}(\sqrt{d})$ is a lattice.*

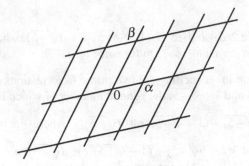

Figure 11.8: A lattice in $\mathbb{C}$.

Proof. Suppose I is a nonzero ideal in the integers of $\mathbb{Q}(\sqrt{d})$, and let α be a nonzero element of I nearest to 0. Since I is closed under sums and also under multiplication by -1, I also includes all ordinary integer multiples of α. But this is not all, because I also includes $\alpha\sqrt{d}$, which is in a direction from 0 perpendicular to the direction of α (since $d < 0$).

Now let $\beta \in I$ be as close as possible to 0 but not an ordinary integer multiple of α. I claim that the lattice $\{\alpha m + \beta n : m, n \in \mathbb{Z}\} = I$.

If not, let γ be a member of I not in $\{\alpha m + \beta n : m, n \in \mathbb{Z}\}$, and consider the parallelogram of the lattice that includes γ (Figure 11.9). Now γ

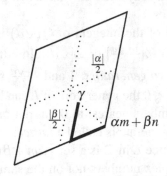

Figure 11.9: Lattice point $\alpha m + \beta n$ nearest to γ.

necessarily lies in one quarter of the parallelogram, in which case its distance from the nearest corner $\alpha m + \beta n$ is less than $\frac{|\alpha|}{2} + \frac{|\beta|}{2}$ by the triangle inequality, hence less than the length $\max(|\alpha|, |\beta|)$ of the longest side of the parallelogram. But then the element $\gamma - (\alpha m + \beta n)$ of I is at distance $< \max(|\alpha|, |\beta|)$ from 0, contrary to the choice of α and β.   □

The proof above does not assume that I is an ideal, only that it is closed under $+$ and $-$ (and hence is a group). Assuming $I = \{\alpha m + \beta n : m, n \in \mathbb{Z}\}$ is an ideal leads to the stronger conclusion that $I = (\alpha) + (\beta)$, the sum of the principal ideals generated by α and β.

We always have $I = \{\alpha m + \beta n : m, n \in \mathbb{Z}\} \subseteq (\alpha) + (\beta)$, since each $\alpha m \in (\alpha)$ and each $\beta n \in (\beta)$. Conversely, $\{\alpha m + \beta n : m, n \in \mathbb{Z}\}$ certainly includes α. Hence if I is an ideal then it includes all members of (α), by closure under multiplication by ring members. Similarly, I includes all members of (β), hence $I \supseteq (\alpha) + (\beta)$ by closure under sums.

Exercises

It should be stressed that, while an ideal $\{\alpha m + \beta n : m, n \in \mathbb{Z}\}$ is necessarily the sum ideal $(\alpha) + (\beta)$, the converse does not always hold. Certainly, $(\alpha) + (\beta) \supseteq \{\alpha m + \beta n : m, n \in \mathbb{Z}\}$, as we observed above, but $(\alpha) + (\beta)$ may include members not of the form $\alpha m + \beta n$ for $m, n \in \mathbb{Z}$. In this case the pair α, β is not an integral basis for $(\alpha) + (\beta)$.

11.6.1 Show that $(5) + (1 + \sqrt{-5})$ in $\mathbb{Z}[\sqrt{-5}]$ includes the element $\sqrt{-5}$.

11.6.2 Show that $\sqrt{-5} \neq 5m + (1 + \sqrt{-5})n$ for all $m, n \in \mathbb{Z}$.

Thus $(5) + (1 + \sqrt{-5}) \neq \{5m + (1 + \sqrt{-5})n : m, n \in \mathbb{Z}\}$, and therefore 5, $1 + \sqrt{-5}$ is not an integral basis for $(5) + (1 + \sqrt{-5})$. However, we know that $(5) + (1 + \sqrt{-5})$ has *some* integral basis α, β, by the lattice property of ideals proved above.

11.6.3 Find $\alpha, \beta \in \mathbb{Z}[\sqrt{-5}]$ such that $(5) + (1 + \sqrt{-5}) = \{\alpha m + \beta n : m, n \in \mathbb{Z}\}$.

11.7 Products and prime ideals

Since we want a principal ideal (s) of a ring R to behave like the element s of R, the product $(s)(t)$ of principal ideals (s) and (t) should be (st). This means that the product of any member rs of (s) and any member $r't$ of (t) is a member $rr'st$ of $(s)(t)$, and hence so too is the sum of any such products. We use this idea to define the product of any two ideals.

Definition. The *product AB* of ideals A and B in a ring R is

$$AB = \{a_1 b_1 + a_2 b_2 + \cdots + a_k b_k : a_i \in A, b_i \in B\}.$$

It is clear that AB is closed under sums and under products by members of R, hence it is an ideal. We now define *prime ideals* as those that have the "prime divisor property" with respect to products of ideals.

Definition. An ideal P is *prime* if, whenever P contains (that is, "divides") the product AB of ideals, P contains A or P contains B.

A more concise way to state this definition of prime ideal is:

$$AB \subseteq P \Rightarrow A \subseteq P \text{ or } B \subseteq P. \tag{1}$$

A definition involving membership rather than containment is:

$$ab \in P \Rightarrow a \in P \text{ or } b \in P, \tag{2}$$

and the two are equivalent by the following theorem.

Equivalent definitions of prime ideal. *The following properties of an ideal P are equivalent:*
  (1) $AB \subseteq P \Rightarrow A \subseteq P$ or $B \subseteq P$,
  (2) $ab \in P \Rightarrow a \in P$ or $b \in P$.

Proof. $(1) \Rightarrow (2)$:

$$ab \in P \Rightarrow (ab) \subseteq P \quad \text{by closure of } P$$
$$\Rightarrow (a)(b) \subseteq P \quad \text{by definition of } (a)(b)$$
$$\Rightarrow (a) \subseteq P \text{ or } (b) \subseteq P \quad \text{by property (1)}$$
$$\Rightarrow a \in P \text{ or } b \in P \quad \text{since } a \in (a), b \in (b).$$

$(2) \Rightarrow (1)$:
Suppose $AB \subseteq P$ and $A \nsubseteq P$, so we want to show that $B \subseteq P$. Since $A \nsubseteq P$, there is an $a \in A$ with $a \notin P$. Then

$$AB \subseteq P \Rightarrow ab \in P \quad \text{for any } b \in B, \text{ by definition of } AB$$
$$\Rightarrow a \in P \text{ or } b \in P \quad \text{for any } b \in B, \text{ by property (2)}$$
$$\Rightarrow b \in P \quad \text{since } a \notin P \text{ by assumption}$$
$$\Rightarrow B \subseteq P \quad \text{since this is for any } b \in B. \qquad \square$$

As mentioned in Section 11.2, prime principal ideals are "maximal" in the sense that the only ideal properly containing them is the whole ring. However, "prime" and "maximal" are not always the same, thus we need a separate definition of maximal ideals.

Definition. An ideal M in a ring R is *maximal* if $M \neq R$ but the only ideals containing M are R and M itself.

We can now prove a relationship between prime and maximal ideals in one direction:

Primality of maximal ideals. *Every maximal ideal is prime.*

Proof. Suppose that M is a maximal ideal, $ab \in M$ and $a \notin M$. Thus (using the second definition of prime ideal) we want to prove that $b \in M$.

Since M is maximal and $a \notin M$, the ideal

$$M[a] = \{ar + ms : r, s \in R, m \in M\},$$

which contains both M and a, must be all of R. This means that 1 in particular is of the form $ar + ms$. Now we can use a familiar trick:

$$1 = ar + ms \Rightarrow b = abr + mbs$$
$$\Rightarrow b \in M, \quad \text{since } ab \in M \text{ and } m \in M. \qquad \square$$

Examples of prime ideals in $\mathbb{Z}[\sqrt{-5}]$

- In Section 11.5 we found that the nonprincipal ideal $(2) + (1 + \sqrt{-5})$ is maximal, hence it is prime by the above theorem.

- Another one is $(3) + (1 + \sqrt{-5}) = \{3m + (1 + \sqrt{-5})n : m, n \in \mathbb{Z}\}$, which is maximal for the following reason. Elements *not* in it are of the form $3m' + 1 + (1 + \sqrt{-5})n'$ or $3m' + 2 + (1 + \sqrt{-5})n'$. Now an ideal containing $(3) + (1 + \sqrt{-5})$ and some $3m' + 1 + (1 + \sqrt{-5})n'$ includes 1, so it is $\mathbb{Z}[\sqrt{-5}]$. An ideal containing $(3) + (1 + \sqrt{-5})$ and some $3m' + 2 + (1 + \sqrt{-5})n'$ includes 2, hence it also includes $1 = 3 - 2$, so this ideal too is $\mathbb{Z}[\sqrt{-5}]$.

- The ideal $(3) + (1 - \sqrt{-5})$ is maximal, hence prime, by a similar argument. It is called the *conjugate* of $(3) + (1 + \sqrt{-5})$, since its members are the conjugates of the members of $(3) + (1 + \sqrt{-5})$.

Like $(2) + (1 + \sqrt{-5})$, $(3) + (1 + \sqrt{-5})$ and its conjugate are non-principal ideals in $\mathbb{Z}[\sqrt{-5}]$. This can be seen by making a picture of $(3) + (1 + \sqrt{-5})$ and observing that no element has neighbors in perpendicular directions, hence it is not the same shape as $\mathbb{Z}[\sqrt{-5}]$. Rather surprisingly, it has the same shape as $(2) + (1 + \sqrt{-5})$ (exercises).

Exercises

In studying the shape of lattices, the key fact to be aware of is that if α and β are complex numbers in different directions from 0, then the ratio of their distances from 0 is $|\alpha/\beta|$ and the angle between their directions is $\arg \alpha/\beta$. Thus the shape of the parallelogram determined by 0, α, and β is determined by the quotient α/β.

11.7.1 Sketch a picture of $(3)+(1+\sqrt{-5})$ accurate enough to show that it is not the same shape as $\mathbb{Z}[\sqrt{-5}]$.

11.7.2 Explain why the lattices $(3)+(1+\sqrt{-5})$ and $(3)+(1-\sqrt{-5})$ have the same shape.

11.7.3 By considering the quotients $2/(1+\sqrt{-5})$ and $(1-\sqrt{-5})/3$, show that the lattices $(2)+(1+\sqrt{-5})$ and $(3)+(1-\sqrt{-5})$ have the same shape.

It turns out that all nonprincipal ideals of $\mathbb{Z}[\sqrt{-5}]$ have the same shape as $(2)+(1+\sqrt{-5})$ (see Section 12.7). Thus exactly two shapes of ideals occur for $\mathbb{Z}[\sqrt{-5}]$. The classical way to say this is that the *class number* of $\mathbb{Z}[\sqrt{-5}]$ is 2.

11.8 Ideal prime factorization

Now that we have a definition of product for ideals, we can define divisibility as we do for any commutative product: *B divides A* means there is a C such that $A = BC$. But we previously suggested that "divides" should mean "contains" for ideals, so now is our chance to test the merit of containment as a concept of divisibility.

Our examples are nonprincipal ideals in $\mathbb{Z}[\sqrt{-5}]$, such as the ideal $(2)+(1+\sqrt{-5})$. To make products easier to read, we write the ideal with integral basis α, β as (α,β); for example $(2)+(1+\sqrt{-5}) = (2,1+\sqrt{-5})$. (This conflicts with the notation for ordered pairs, however no ordered pairs will turn up to cause confusion.)

The first example is the prime ideal $(2,1+\sqrt{-5})$, which contains the principal ideal (2). Is there an ideal C such that

$$(2) = (2,1+\sqrt{-5})C?$$

Happily, the answer is yes, and in fact we have the following.

Ideal prime factorization of (2). $(2) = (2,1+\sqrt{-5})^2$.

Proof. It follows from the definition of product of ideals that

$$4 = 2 \times 2 \in (2, 1 + \sqrt{-5})^2,$$
$$2 + 2\sqrt{-5} = 2 \times (1 + \sqrt{-5}) \in (2, 1 + \sqrt{-5})^2,$$
$$-4 + 2\sqrt{-5} = (1 + \sqrt{-5})^2 \in (2, 1 + \sqrt{-5})^2.$$

And

$$4, 2 + 2\sqrt{-5}, -4 + 2\sqrt{-5} \in (2, 1 + \sqrt{-5})^2$$
$$\Rightarrow 2 \in (2, 1 + \sqrt{-5})^2 \quad \text{by closure under } + \text{ and } -$$
$$\Rightarrow \text{ all multiples of } 2 \in (2, 1 + \sqrt{-5})^2$$
$$\text{by closure under multiplication by members of } \mathbb{Z}[\sqrt{-5}]$$
$$\Rightarrow (2) \subseteq (2, 1 + \sqrt{-5})^2.$$

Conversely, any element of $(2, 1 + \sqrt{-5})^2$ is a sum of products of terms $2m$ and $(1 + \sqrt{-5})n$. Any product involving $2m$ is a multiple of 2, and so is any product involving $(1 + \sqrt{-5})^2 = -4 + 2\sqrt{-5}$. Therefore, any element of $(2, 1 + \sqrt{-5})^2$ is a multiple of 2, hence $(2, 1 + \sqrt{-5})^2 \subseteq (2)$, as required. $\square$

Likewise, the two prime ideals $(3, 1 + \sqrt{-5})$ and $(3, 1 - \sqrt{-5})$ contain (3), and they are in fact its ideal prime factors.

Ideal prime factorization of (3)**.** $(3) = (3, 1 + \sqrt{-5})(3, 1 - \sqrt{-5})$.

Proof. It follows from the definition of product of ideals that

$$9 = 3 \times 3 \in (3, 1 + \sqrt{-5})(3, 1 - \sqrt{-5}),$$
$$6 = (1 + \sqrt{-5})(1 - \sqrt{-5}) \in (3, 1 + \sqrt{-5})(3, 1 - \sqrt{-5}).$$

And

$$9, 6 \in (3, 1 + \sqrt{-5})(3, 1 - \sqrt{-5})$$
$$\Rightarrow 3 \in (3, 1 + \sqrt{-5})(3, 1 - \sqrt{-5}) \quad \text{by closure under } -$$
$$\Rightarrow \text{ all multiples of } 3 \in (3, 1 + \sqrt{-5})(3, 1 - \sqrt{-5})$$
$$\text{by closure under multiplication by members of } \mathbb{Z}[\sqrt{-5}]$$
$$\Rightarrow (3) \subseteq (3, 1 + \sqrt{-5})(3, 1 - \sqrt{-5}).$$

Conversely, any element of $(3,1+\sqrt{-5})(3,1-\sqrt{-5})$ is a sum of products of terms $3m$ and $(1\pm\sqrt{-5})n$. Any product involving $3m$ is a multiple of 3, and so is any product involving $(1+\sqrt{-5})(1-\sqrt{-5}) = 6$. It follows that any element of $(3,1+\sqrt{-5})(3,1-\sqrt{-5})$ is a multiple of 3, hence $(3,1+\sqrt{-5})(3,1-\sqrt{-5}) \subseteq (3)$ as required. $\square$

These two factorizations imply that the prime factorization 2×3 of 6 in $\mathbb{Z}[\sqrt{-5}]$ actually splits further, into the *ideal prime factorization*

$$6 = (2,1+\sqrt{-5})^2(3,1+\sqrt{-5})(3,1-\sqrt{-5}),$$

when we replace 2 and 3 by the corresponding principal ideals (2) and (3). Even more marvellous, the ideal factors recombine to produce the *other* prime factorization in $\mathbb{Z}[\sqrt{-5}]$, $6 = (1+\sqrt{-5})(1-\sqrt{-5})$ (where $(1+\sqrt{-5})$ and $(1-\sqrt{-5})$ are taken as principal ideals), because

$$(1+\sqrt{-5}) = (2,1+\sqrt{-5})(3,1+\sqrt{-5}),$$
$$(1-\sqrt{-5}) = (2,1+\sqrt{-5})(3,1-\sqrt{-5}).$$

The latter factorizations can be checked along the same lines as the ideal prime factorizations of (2) and (3) above (exercises).

Thus the two prime factorizations 2×3 and $(1+\sqrt{-5})(1-\sqrt{-5})$ of 6 are actually different groupings of factors in the *same* ideal prime factorization. Admittedly, this does not prove that ideal prime factorization is unique in $\mathbb{Z}[\sqrt{-5}]$, but it shows how uniqueness might be possible, and the next chapter will explain why it is in fact true.

Exercises

11.8.1 Show that $1+\sqrt{-5} \in (2,1+\sqrt{-5})(3,1+\sqrt{-5})$, so that

$$(1+\sqrt{-5}) \subseteq (2,1+\sqrt{-5})(3,1+\sqrt{-5}).$$

11.8.2 Show that $1+\sqrt{-5}$ divides each element of $(2,1+\sqrt{-5})(3,1+\sqrt{-5})$. Hence deduce from Exercise 11.8.1 that

$$(1+\sqrt{-5}) = (2,1+\sqrt{-5})(3,1+\sqrt{-5}).$$

11.8.3 Show similarly that $(1-\sqrt{-5}) = (2,1-\sqrt{-5})(3,1-\sqrt{-5})$.

11.9 Discussion

The failure of unique prime factorization was a deeply hidden problem, which remained undetected for nearly two centuries after its side effects were first noticed. The first such effect is the anomalous behavior of the quadratic form $x^2 + 5y^2$, noticed by Fermat in 1654. As we know, Fermat had successfully classified primes of the form $x^2 + y^2$, $x^2 + 2y^2$, and $x^2 + 3y^2$ before this, and we have seen in Chapters 7 and 9 how his classification can be explained with the help of unique prime factorization in $\mathbb{Z}[\sqrt{-1}]$, $\mathbb{Z}[\sqrt{-2}]$, and $\mathbb{Z}[\frac{-1+\sqrt{-3}}{2}]$ respectively. Fermat presumably did *not* conceptualize the problem this way, otherwise he would have seen trouble in store for $x^2 + 5y^2$, due the failure of unique prime factorization in $\mathbb{Z}[\sqrt{-5}]$.

He did indeed have trouble, but for reasons that are unclear he left only a conjecture about prime *products* of the form $x^2 + 5y^2$: *if p_1, p_2 are primes of the form $20n + 3$ or $20n + 7$, then $p_1 p_2 = x^2 + 5y^2$.*

This is strange, because primes of the form $x^2 + 5y^2$ exist (for example $29 = 3^2 + 5 \times 2^2$) and an easy application of congruences mod 20 shows that any prime $p = x^2 + 5y^2$ must be of the form $20n + 1$ or $20n + 9$. Euler (1744) evidently realized this, and conjectured that the converse is also true, thus:

$$p = x^2 + 5y^2 \Leftrightarrow p = 20n + 1 \text{ or } 20n + 9.$$

Putting the conjectures of Fermat and Euler together, we have the conjecture that $x^2 + 5y^2$ is the form of

- primes of the form $20n + 1$ or $20n + 9$,

- products of two primes of the form $20n + 3$ or $20n + 7$.

The first to account for this two-faced behavior of $x^2 + 5y^2$ was Lagrange (1773), who discovered that $x^2 + 5y^2$ has a hidden companion—the quadratic form $2x^2 + 2xy + 3y^2$—whose prime values are of the form $20n + 3$ or $20n + 7$. Lagrange made this discovery through his theory of *equivalence of quadratic forms* that we introduced in Section 5.6. He discovered that equivalent forms have the same determinant, and that:

> *All forms with determinant 1 are equivalent to $x^2 + y^2$.*
> *All forms with determinant 2 are equivalent to $x^2 + 2y^2$.*
> *All forms with determinant 3 are equivalent to $x^2 + 3y^2$.*

Whereas:

For determinant 5 there are two inequivalent forms: $x^2 + 5y^2$ *and*
$$2x^2 + 2xy + 3y^2.$$

These discoveries throw new light on the regular behavior of $x^2 + y^2$, $x^2 + 2y^2$, and $x^2 + 3y^2$, and also suggest why $x^2 + 5y^2$ is irregular. The determinants 1, 2, and 3 each have only one equivalence class of forms, or *class number 1*, which is the simplest situation (now recognized as equivalent to unique prime factorization in the corresponding ring). The determinant 5 has two equivalence classes, or *class number 2*, and this is more complicated because the two forms interact with each other. Lagrange saw that the product of two numbers of the form $2x^2 + 2xy + 3y^2$ is of the form $x^2 + 5y^2$, because

$$(2x_1^2 + 2x_1y_1 + 3y_1^2)(2x_2^2 + 2x_2y_2 + 3y_2^2) = X^2 + 5Y^2,$$

where $X = 2x_1x_2 + x_1y_2 + x_2y_1 - 2y_1y_2$ and $Y = x_1y_2 + x_2y_1 + y_1y_2$. He also saw that a number of the form $x^2 + 5y^2$ times a number of the form $2x^2 + 2xy + 3y^2$ is again of the form $2x^2 + 2xy + 3y^2$ because

$$(x_1^2 + 5y_1^2)(2x_2^2 + 2x_2y_2 + 3y_2^2) = 2X^2 + 2XY + 3Y^2,$$

where $X = x_1x_2 - y_1x_2 - 3y_1y_2$ and $Y = x_1y_2 + 2y_1x_2 + y_1y_2$.

These stellar feats of high school algebra can be emulated in a fairly mechanical way by using factorizations in $\mathbb{Q}[\sqrt{-5}]$ and combining terms so as to produce conjugate factors. For example, since

$$x^2 + 5y^2 = (x + y\sqrt{-5})(x - y\sqrt{-5}),$$
$$2x^2 + 2xy + 3y^2 = 2\left[x + \frac{y}{2}(1 + \sqrt{-5})\right]\left[x + \frac{y}{2}(1 - \sqrt{-5})\right]$$

we have

$$(x_1^2 + 5y_1^2)(2x_2^2 + 2x_2y_2 + 3y_2^2)$$
$$= (x_1 + y_1\sqrt{-5})(x_1 - y_1\sqrt{-5})2\left[x_2 + \frac{y_2}{2}(1 + \sqrt{-5})\right]\left[x_2 + \frac{y_2}{2}(1 - \sqrt{-5})\right]$$
$$= 2(x_1 + y_1\sqrt{-5})\left[x_2 + \frac{y_2}{2}(1 + \sqrt{-5})\right](x_1 - y_1\sqrt{-5})\left[x_2 + \frac{y_2}{2}(1 - \sqrt{-5})\right]$$
$$= 2\left[x_1x_2 - y_1x_2 - 3y_1y_2 + \frac{x_1y_2 + 2y_1x_2 + y_1y_2}{2}(1 + \sqrt{-5})\right] \times \text{its conjugate}$$
$$= 2X^2 + 2XY + 3Y^2,$$

where $X = x_1x_2 - y_1x_2 - 3y_1y_2$, $Y = x_1y_2 + 2y_1x_2 + y_1y_2$.

But does something similar happen for any determinant?

Lagrange's results on the two inequivalent forms with determinant 5 were early steps in a theory later known as *composition of quadratic forms*. The first steps were Diophantus' identity and Brahmagupta's generalization of it:

$$(x_1^2 - ny_1^2)(x_2^2 - ny_2^2) = X^2 - nY^2,$$

where $X = x_1 x_2 + n y_1 y_2$ and $Y = x_1 y_2 + y_1 x_2$. The Brahmagupta identity says that "the form $x^2 - ny^2$, composed with itself, yields itself". If we denote the form $x^2 + 5y^2$ by A and the form $2x^2 + 2xy + 3y^2$ by B, then Lagrange's results (combined with Brahmagupta's) say that the composites of A and B have the following "multiplication table":

$$A^2 = A, \quad AB = BA = B, \quad B^2 = A.$$

We recognize this as the multiplication table for the two-element group with identity element A. Today it is called the *class group* for $\mathbb{Q}(\sqrt{-5})$ and it is defined in an entirely different way, using ideals.

The classes A, B of the inequivalent forms $x^2 + 5y^2$, $2x^2 + 2xy + 3y^2$ correspond to two classes of ideals of $\mathbb{Z}[\sqrt{-5}]$: the class A^* of principal ideals and the class B^* of nonprincipal ideals (in this ring, all nonprincipal ideals are equivalent in the sense of having the same shape). The products of ideals computed in the previous section show that these ideal classes A^* and B^* have the same multiplication table as the forms A and B.

I think it will be agreed that it is easier to multiply ideals than to compose forms, but it is not easy to see why they are really the same thing. For this reason, replacement of forms by ideals was a momentous change of direction for number theory. Early work in the direction of ideals, such as Euler's use of quadratic integers, was snuffed out when Gauss apparently realized that unique prime factorization could fail. His way around this obstacle was to develop Lagrange's theory of quadratic forms for arbitrary determinant, which he did, in stupefying complexity, in his *Disquisitiones* of 1801.

Gauss took a small step toward a rigorous theory of quadratic integers with his proof, published in 1832, of unique factorization in $\mathbb{Z}[i]$. This effectively made the theory of the form $x^2 + y^2$ obsolete. However, it was only in the 1840s and 1850s that Dirichlet, Kummer, Kronecker, and Dedekind began developing general alternatives to composition of forms. As mentioned earlier, Kummer had the idea of "ideal numbers", and he used it successfully in the theory of cyclotomic integers, where there was no viable alternative. The easier theory of quadratic integers did not emerge until later, perhaps because Dirichlet and Dedekind had invested a lot of time in simplifying the competing theory of composition of forms.

Composition of forms began to fade only in the 1870s, when Dedekind developed ideal theory for algebraic integers of arbitrary degree. In 1877 he gave a careful exposition of the $\mathbb{Z}[\sqrt{-5}]$ example to motivate his general theory of ideals of algebraic integers. His very readable little book, in English translation as Dedekind (1877), is recommended for its insight into Dedekind's struggle to make "ideal numbers" actual. Also recommended is the translator's introduction, which discusses the historical steps from quadratic forms to algebraic integers in rather more detail than is possible here.

12

Prime ideals

PREVIEW

In this final chapter we regain unique prime factorization in rings such as $\mathbb{Z}[\sqrt{-5}]$ by using prime *ideals* instead of prime *numbers*.

We first note the pure algebraic meaning of ideals in a ring R: ideals are the subsets I for which the notion of "congruence mod I" makes sense. This generalizes the idea of congruence mod n in $\mathbb{Z}$, and there is a corresponding generalization of the ring $\mathbb{Z}/n\mathbb{Z}$, namely the *quotient ring R/I*.

Properties of the ideal I (in particular, being prime or maximal) are reflected in properties of the quotient ring R/I (being an integral domain or field respectively). For the integers of an imaginary quadratic field, "prime" turns out to be equivalent to "maximal", which helps to prove the key property of ideals: that *"contains" means "divides"*: $B \supseteq A \Rightarrow A = BC$ for some ideal C.

The first step is to introduce the *conjugate $\overline{A}$* of an ideal A, and to prove that $A\overline{A}$ is a principal ideal. This is used to boost results about principal ideals (which are easy because principal ideals behave like numbers) to results about general ideals.

We use this strategy to prove that "contains" means "divides" and to obtain unique factorization of ideals into prime ideals.

Returning finally to the specific case of $\mathbb{Z}[\sqrt{-5}]$, we take a brief look at the concept of *ideal classes*, in order to show that all nonprincipal ideals of $\mathbb{Z}[\sqrt{-5}]$ have the same shape. We need this result to conclude our unfinished business with $\mathbb{Z}[\sqrt{-5}]$—the classification of the primes of the form $x^2 + 5y^2$.

12.1   Ideals and congruence

We now know that ideals can serve as "ideal numbers" in situations where actual numbers seem to be lacking, for example in $\mathbb{Z}[\sqrt{-5}]$, where 2 and $1 + \sqrt{-5}$ should have a gcd $\neq 1$ but don't. However, ideals also have a natural abstract function: *an ideal I in a ring R is a subset of R for which "congruence mod I" makes sense.*

Given an ideal I, we define congruence mod I by

$$a \equiv b \pmod{I} \quad \Leftrightarrow \quad a - b \in I.$$

Then the equivalence properties of $\equiv$ follow from closure properties of I:

- $a \equiv a \pmod{I}$
  because $a \in I \Rightarrow -a \in I$, since multiples of -1 are in I, which in turn implies $a + (-a) = 0 \in I$ by closure under $+$.

- $a \equiv b \pmod{I} \Rightarrow b \equiv a \pmod{I}$
  because $a - b \in I \Rightarrow b - a \in I$, by closure under multiplying by -1 again.

- $a \equiv b \pmod{I}$ and $b \equiv c \pmod{I} \Rightarrow a \equiv c \pmod{I}$
  because $a - b \in I$ and $b - c \in I \Rightarrow (a - b) + (b - c) = a - c \in I$, by closure under $+$.

It follows that R is partitioned into *congruence classes* $I + a$, where

$$I + a = \{i + a : i \in I\}.$$

Moreover, it is meaningful to add and multiply classes by the rules

$$(I + a) + (I + b) = I + (a + b),$$
$$(I + a)(I + b) = I + ab.$$

This can be proved in exactly the same way as in Section 3.2 for congruence classes $n\mathbb{Z} + a$ and $n\mathbb{Z} + b$ in $\mathbb{Z}$.

Any element of $I + a$ is of the form $k + a$ for some $k \in I$, so if we add it to an arbitrary element $l + b$ of $I + b$, where $l \in I$, we get

$$(k + l) + (a + b),$$

which is in $I + (a + b)$ because $k + l = i \in I$ by closure under sums.

If we multiply the element $k + a \in I + a$ by $l + b \in I + b$ we get

$$kl + kb + la + ab,$$

which is in $I + ab$ because

$$k, l \in I \Rightarrow kl, kb, la \in I \quad \text{by closure under products by members of } R$$
$$\Rightarrow kl + kb + la \in I \quad \text{by closure under sums.}$$

(It should be mentioned at this point that we are assuming R to be commutative, as is the case for the number rings we are interested in. For noncommutative rings one has to distinguish between *left* and *right* ideals.)

Finally, the set R/I of congruence classes, under the $+$ and $\times$ operations just defined, inherits the ring properties from R. For example, multiplication is commutative in R/I because it is in R:

$$(I + a)(I + b) = I + ab \quad \text{by definition of } \times \text{ in } R/I$$
$$= I + ba \quad \text{by commutativity in } R$$
$$= (I + b)(I + a) \quad \text{by definition of } \times \text{ in } R/I.$$

The other properties can be checked similarly, and hence R/I is a ring, called the *quotient ring* of R by the ideal I.

If one pursues the study of R/I following the model of $\mathbb{Z}/n\mathbb{Z}$ in Chapter 3, then the next question is: for which ideals I is R/I a field? In $\mathbb{Z}$, this happens when n is prime, but in general rings the answer is not so simple. We take up the question in the next section.

Exercises

12.1.1 Which congruence classes play the roles of 1 and 0 in R/I?

12.1.2 Check the other ring properties of R/I.

It is useful to visualize the congruence classes for some actual ideals I, say in $\mathbb{Z}[\sqrt{-5}]$.

12.1.3 Pick out the congruence classes of $I = (2) + (1 + \sqrt{-5})$ in Figure 11.7.

12.1.4 Sketch a picture of $I = (3) + (1 + \sqrt{-5})$ in $\mathbb{Z}[\sqrt{-5}]$ and show that it has three congruence classes.

12.2   Prime and maximal ideals

In Section 11.7 we saw that maximal ideals are prime. However, prime ideals are not always maximal, and the difference between the two is nicely captured by properties of the quotient ring R/I of R by the ideal I.

Characterization of prime ideals. I *is a prime ideal of a ring* $R \Leftrightarrow R/I$ *has no zero divisors. (Zero divisors are nonzero congruence classes* $I + a$, $I + b$ *whose product* $I + ab$ *is* I, *the class of 0.)*

Proof. ($\Rightarrow$) Suppose I is prime, so we have to prove that R/I has no zero divisors.

$$I + ab \text{ is the class of } 0 \Rightarrow ab \in I$$
$$\Rightarrow a \in I \text{ or } b \in I \quad \text{since } I \text{ is prime}$$
$$\Rightarrow I + a \text{ or } I + b \quad \text{is the class of } 0$$
$$\Rightarrow R/I \text{ has no zero divisors.}$$

($\Leftarrow$) Suppose R/I has no zero divisors, so we have to prove that I is a prime ideal.

$$ab \in I \Rightarrow I + ab = I \quad \text{(the class of 0)}$$
$$\Rightarrow (I + a)(I + b) = I \quad \text{by definition of product of congruence classes}$$
$$\Rightarrow I + a = I \text{ or } I + b = I \quad \text{since } R/I \text{ has no zero divisors}$$
$$\Rightarrow a \in I \text{ or } b \in I$$
$$\Rightarrow I \text{ is a prime ideal.} \qquad \square$$

Characterization of maximal ideals. I *is a maximal ideal of a ring* $R \Leftrightarrow R/I$ *is a field. (That is, each nonzero member of* R/I *has a multiplicative inverse.)*

Proof. ($\Rightarrow$) Suppose I is maximal, so we have to prove that R/I is a field.

$$I + a \text{ a nonzero congruence class} \Rightarrow a \notin I$$
$$\Rightarrow \text{ideal } \{ir + as : r, s \in R, i \in I\} = R$$
$$\text{by maximality of } I$$
$$\Rightarrow 1 = ir + as \quad \text{for some } r, s \in R, i \in I$$
$$\Rightarrow I + as = I + 1$$
$$\Rightarrow I + a \text{ has inverse } I + s$$
$$\Rightarrow R/I \text{ is a field.}$$

($\Leftarrow$) Suppose each nonzero congruence class $I + a$ has an inverse $I + s$, so we have to prove that I is maximal, that is, the only ideal containing I and an $a \notin I$ is R.

$a \notin I \Rightarrow I + a$ is a nonzero congruence class
$$\Rightarrow I + as = I + 1 \text{ for some } s \in R,$$
namely, for the inverse class $I + s$ of $I + a$
$$\Rightarrow 1 = ir + as \text{ for some } r \in R \text{ and } i \in I$$
$\Rightarrow$ any ideal containing I and a includes 1 and hence equals R
$\Rightarrow I$ is maximal. $\square$

Remark. A ring with no zero divisors is called an *integral domain*. Any field is an integral domain, but an integral domain is not necessarily a field. For example, $\mathbb{Z}$ is an integral domain but not a field. However, an integral domain always has the following *cancellation property* in common with fields:
$$ab = ac \text{ and } a \neq 0 \Rightarrow b = c.$$
This is because
$$ab = ac \Rightarrow ab - ac = 0$$
$$\Rightarrow a(b - c) = 0$$
$$\Rightarrow b - c = 0 \quad \text{since } a \neq 0 \text{ and there are no zero divisors}$$
$$\Rightarrow b = c.$$

Exercises

A non-maximal prime ideal is easy to find in $\mathbb{Z}[\sqrt{-5}]$; for example, the principal ideal (2).

12.2.1 Find the three nonzero elements of $\mathbb{Z}[\sqrt{-5}]/(2)$ and show that they are not zero divisors, hence (2) is a prime ideal.

12.2.2 Why is (2) not maximal in $\mathbb{Z}[\sqrt{-5}]$?

12.3   Prime ideals of imaginary quadratic fields

There is generally a big difference between integral domains and fields, and a correspondingly big difference between prime and maximal ideals.

However, there is one important case where integral domains are always fields, namely, when they are finite.

Lemma. *A finite integral domain is a field.*

Proof. Let a be a nonzero element of D, and consider $a^2, a^3, a^4, \ldots$ Since D is finite, some value a^m in this sequence recurs as a later value a^{m+n}, so

$$a^{m+n} = a^m, \text{ and hence } a^n = 1 \text{ by cancellation.}$$

But this means $a \cdot a^{n-1} = 1$, so a has the multiplicative inverse a^{n-1} and hence D is a field. $\square$

In view of this lemma, and the characterizations of prime and maximal ideals in Section 12.2, a prime ideal in a ring R will be maximal whenever R/I is finite. This leads to the following theorem.

Maximality of prime ideals in imaginary quadratic fields. *A prime ideal in the integers of an imaginary quadratic field is maximal.*

Proof. Let R be the ring of integers of the quadratic field and I a prime ideal. By the results above, it suffices to show that R/I is finite; in other words, that there are finitely many congruence classes $I + r$ as r ranges over R. But this is clear from the lattice property of ideals in Section 11.6. The congruence classes $I + r$ are represented by the r in *one* parallelogram of the lattice I (Figure 12.1) because an r' in any other parallelogram of I is congruent (mod I) to such an r.

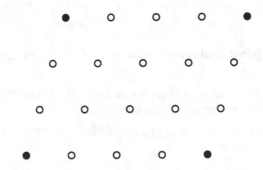

Figure 12.1: One parallelogram of the lattice I.

Thus there are only finitely many congruence classes $I + r$. $\square$

Exercises

We have restricted attention to imaginary quadratic fields because their integers
are easy to visualize and most of our motivating examples lead to them. A ring
of real quadratic integers, such as $\mathbb{Z}[\sqrt{2}]$, is not a "lattice" in the literal sense
because it lies densely along the real axis. However, it is obvious that $\mathbb{Z}[\sqrt{2}]$ has
the integral basis $1, \sqrt{2}$ and we can prove that its ideals have integral bases too.

The idea is to map elements $a + b\sqrt{2}$ of $\mathbb{Z}[\sqrt{2}]$ to points $a + bi\sqrt{2}$ of $\mathbb{C}$. Then
we can apply geometric arguments to the image points.

12.3.1 Show that any subset I of $\mathbb{Z}[\sqrt{2}]$ closed under $+$ and $-$ is thereby mapped
to a subset I^* of $\mathbb{C}$ with preservation of sums and differences. If I is an
ideal, show that I^* does not lie in a line.

12.3.2 Deduce from Exercise 12.3.1 and the lattice theorem of Section 11.6 that
I has an integral basis.

The same argument obviously applies to the integers of any real quadratic field.
Finally, the parallelogram argument used above can be carried over to these integers.

12.3.3 Show that the map $a + b\sqrt{2} \mapsto a + bi\sqrt{2}$ sends cosets of I to cosets of I^*,
and deduce that $\mathbb{Z}[\sqrt{2}]/I$ is finite. (And similarly for the integers of any
real quadratic field.)

12.4 Conjugate ideals

The key to the success of ideal prime factorization in quadratic fields is the
fact, proved below, that every ideal divides a principal ideal. This allows us
to replace questions about ideal factors by easier questions about principal
ideal factors. The trick is to multiply an ideal A by its conjugate $\overline{A}$, the set
of conjugates of elements of A. Just as the product of conjugate quadratic
integers is an ordinary integer k, the product of conjugate ideals turns out
to be the ideal (k) of multiples of an ordinary integer.

Product of conjugate ideals. *If R is the ring of integers of an imaginary
quadratic field and A is an ideal of R, then*

$$A\overline{A} = (k) \quad \text{for some } k \in \mathbb{Z}.$$

Proof. We know from Section 11.6 that $A = \{\alpha m + \beta n : m, n \in \mathbb{Z}\}$ for two
integers α and β of R. Hence $\overline{A} = \{\overline{\alpha} m + \overline{\beta} n : m, n \in \mathbb{Z}\}$ and, by definition
of product of ideals, $A\overline{A} = \{s\alpha\overline{\alpha} + t\beta\overline{\beta} + u\overline{\alpha}\beta + v\alpha\overline{\beta} : s, t, u, v \in \mathbb{Z}\}$.

Now $\alpha\overline{\alpha}$, $\beta\overline{\beta}$, and $\overline{\alpha}\beta + \alpha\overline{\beta}$ are self-conjugate, and hence real. We first find an ordinary integer k that divides them, and then see what follows:

$\alpha\overline{\alpha}, \beta\overline{\beta}, \overline{\alpha}\beta + \alpha\overline{\beta}$ real integers of R

$\Rightarrow \alpha\overline{\alpha}, \beta\overline{\beta}, \overline{\alpha}\beta + \alpha\overline{\beta} \in \mathbb{Z}$ by the nature of integers of R (Section 10.4)

$\Rightarrow \gcd(\alpha\overline{\alpha}, \beta\overline{\beta}, \overline{\alpha}\beta + \alpha\overline{\beta}) = k \in \mathbb{Z}$

$\Rightarrow k = p\alpha\overline{\alpha} + q\beta\overline{\beta} + r(\overline{\alpha}\beta + \alpha\overline{\beta})$ for some $p, q, r \in \mathbb{Z}$

   by the Euclidean algorithm

$\Rightarrow k \in A\overline{A}$

$\Rightarrow (k) \subseteq A\overline{A}$.

Conversely, to show that $(k) \supseteq A\overline{A}$, it suffices to show that k divides the four generators $\alpha\overline{\alpha}$, $\beta\overline{\beta}$, $\overline{\alpha}\beta$, and $\alpha\overline{\beta}$ of $A\overline{A}$. It divides $\alpha\overline{\alpha}$ and $\beta\overline{\beta}$ by construction, and it divides $\alpha\overline{\beta}$ and $\overline{\alpha}\beta$ provided $(\alpha\overline{\beta})/k$ and $(\overline{\alpha}\beta)/k$ belong to R. The latter numbers are the roots of

$$\left(x - \frac{\alpha\overline{\beta}}{k}\right)\left(x - \frac{\overline{\alpha}\beta}{k}\right) = x^2 - \frac{\alpha\overline{\beta} + \overline{\alpha}\beta}{k}x + \frac{\alpha\overline{\alpha}}{k} \cdot \frac{\beta\overline{\beta}}{k} = 0,$$

the coefficients of which we know to be ordinary integers. Hence $(\alpha\overline{\beta})/k$ and $(\overline{\alpha}\beta)/k$ are quadratic integers, and hence members of R. $\square$

Exercises

The proof above assumes the lattice property of ideals, or rather that each ideal has an integral basis α, β, and we know from the previous exercise set that this is also true for rings of integers in real quadratic fields.

12.4.1 Check that the proof above works for real quadratic fields, where the conjugation operation now is $a + b\sqrt{d} \mapsto a - b\sqrt{d}$.

The product of conjugate ideals helps to establish the existence of the *ideal class group*, because it shows the existence of inverses. The identity class of this group is the class of principal ideals, so the theorem says that any ideal (class) times its conjugate (class) is the identity class, hence conjugate ideal classes are inverse to each other.

12.4.2 Bearing in mind that equivalent ideals are the same shape, show that the ideal $(2, 1 + \sqrt{-5})$ is self-conjugate, hence self-inverse.

12.4.3 We already know that $(2, 1 + \sqrt{-5})$ is self-inverse. Why?

12.5   Divisibility and containment

We are now almost ready to prove that "contains means divides" for all ideals in the ring of integers of an imaginary quadratic field. We know that this is the case for principal ideals, and the only other result we need is the following cancellation property.

Cancellation of ideals. *If A, B, C are nonzero ideals of R and $AB \supseteq AC$, then $B \supseteq C$.*

Proof. In the special case where $A = (\alpha)$ is principal

$$AB \supseteq AC \Rightarrow (\alpha)A \supseteq (\alpha)B$$
$$\Rightarrow \alpha B \supseteq \alpha C \quad \text{because the ideals } B \text{ and } C$$
$$\text{are closed under } \times \text{ by multiples of } \alpha,$$
$$\text{and hence } (\alpha)B = \alpha B \text{ and } (\alpha)C = \alpha C$$
$$\Rightarrow B \supseteq C, \quad \text{multiplying both sides by } \alpha^{-1}.$$

In the general case,

$$AB \supseteq AC \Rightarrow \overline{A}AB \supseteq \overline{A}AC, \quad \text{multiplying both sides by } \overline{A}$$
$$\Rightarrow (k)B \supseteq (k)C \quad \text{for some } k, \text{ by Section 12.4}$$
$$\Rightarrow B \supseteq C \quad \text{by the special case.} \qquad \square$$

This is our first application of the trick from Section 12.4—multiplying an ideal by its conjugate to reduce to the easier case of a principal ideal. Cancellation now allows us to extend this trick further: we can multiply an ideal by its conjugate to reduce to the special (and easy) case of a principal ideal, then cancel the conjugate to get back to the general case. This is how we prove that "contains means divides."

"Contains means divides." *If A and B are ideals of R and $B \supseteq A$ then B divides A; that is,*

$$A = BC \quad \text{for some ideal } C.$$

Proof. In the special case where $B = (\beta)$ is a principal ideal

$$B \supseteq A \Rightarrow (\beta) \supseteq A$$
$$\Rightarrow \beta \text{ divides each member of } A$$
$$\Rightarrow A = (\beta)\{\alpha/\beta : \alpha \in A\}$$
$$\Rightarrow A = BC$$

where B is the ideal (β) and $C = \{\alpha/\beta : \alpha \in A\}$ is also an ideal of R. The α/β in C belong to R since each α is divisible by β, and they are closed under $+$ and under multiplication by elements of R since this is true of the α in A.

In the general case

$$B \supseteq A \Rightarrow B\overline{B} \supseteq A\overline{B}, \quad \text{multiplying both sides by } \overline{B}$$
$$\Rightarrow (k) \supseteq A\overline{B} \quad \text{by Section 12.4}$$
$$\Rightarrow A\overline{B} = (k)C \quad \text{for some ideal } C, \text{ by the special case}$$
$$\Rightarrow A\overline{B} = \overline{B}BC \quad \text{since } (k) = \overline{B}B$$
$$\Rightarrow A = BC \quad \text{by cancellation of the ideal } \overline{B}. \qquad \square$$

12.6   Factorization of ideals

We are now ready to prove existence and uniqueness of prime ideal factorization in the ring R of integers of an imaginary quadratic field. Since prime ideals are maximal in this setting, the usual process of finding smaller and smaller factors is replaced by a process of finding larger and larger ideals.

Existence. *Every nonzero ideal $A \neq R$ is a product of prime ideals.*

Proof. If A is not prime, it is not maximal by Section 11.7, hence there is an ideal $B \supsetneq A$ with $B \neq R$. Since "contains means divides," it follows that $A = BC$ for some ideal C.

If B or C is not prime we factorize it similarly, and so on. This process terminates in a finite number of steps (giving a prime factorization) because each nonzero ideal I has only a finite number of congruence classes $I + r$ (Section 12.3) and each extension to a larger ideal absorbs at least one such class. $\qquad \square$

Uniqueness. *The factorization of a nonzero ideal into prime ideals is unique, up to the order of factors.*

Proof. As always, uniqueness follows from existence and a *prime divisor property*, in this case: if a prime ideal P divides a product of ideals AB then P divides A or P divides B.

By definition (Section 11.7) a prime ideal P has the property that if $P \supseteq AB$ then $P \supseteq A$ or $P \supseteq B$. The prime divisor property now follows because "contains means divides." $\qquad \square$

Exercises

According to unique prime ideal factorization, distinct prime *number* factorizations should split into the same prime ideal factorization. We saw this happen in Section 11.8 with the two prime number factorizations of 6 in $\mathbb{Z}[\sqrt{-5}]$. Another example in $\mathbb{Z}[\sqrt{-5}]$ is

$$9 = 3 \times 3 = (2+\sqrt{-5})(2-\sqrt{-5}).$$

12.6.1 Use norms to show that $3, 2+\sqrt{-5}$, and $2-\sqrt{-5}$ are prime numbers of $\mathbb{Z}[\sqrt{-5}]$.

Now we know from Section 11.8 that $(3) = (3, 1+\sqrt{-5})(3, 1-\sqrt{-5})$, so these prime ideal factors of 3 should also be ideal factors $2+\sqrt{-5}$ and $2-\sqrt{-5}$.

12.6.2 Show that $2-\sqrt{-5} \in (3, 1+\sqrt{-5})^2$, so $(2-\sqrt{-5}) \subseteq (3, 1+\sqrt{-5})^2$.

12.6.3 Show, conversely, that $2-\sqrt{-5}$ divides the elements $9, 3+3\sqrt{-5}$, and $-4+2\sqrt{-5}$ generating $(3, 1+\sqrt{-5})^2$, so $(3, 1+\sqrt{-5})^2 \subseteq (2-\sqrt{-5})$.

Thus the factor $2-\sqrt{-5}$ of 9 has prime ideal factorization $(3, 1+\sqrt{-5})^2$, and it remains to show that the other factor $2+\sqrt{-5}$ has ideal prime factorization $(3, 1-\sqrt{-5})^2$.

12.6.4 $2-\sqrt{-5} = (3, 1+\sqrt{-5})^2$ implies $2+\sqrt{-5} = (3, 1-\sqrt{-5})^2$. Why?

12.7   Ideal classes

As an application of unique prime ideal factorization, we determine the primes of the form $x^2 + 5y^2$, thus solving the problem that puzzled Fermat and Euler. To do this we need to know a little more about the ideals of $\mathbb{Z}[\sqrt{-5}]$, namely that they fall into two classes: the principal ideals, all of which have the shape of $\mathbb{Z}[\sqrt{-5}]$, and the nonprincipal ideals, all of which have the shape of $(2, 1+\sqrt{-5}) = \{2m+(1+\sqrt{-5})n : m, n \in \mathbb{Z}\}$. In general, the number of ideal shapes in the ring of integers of an imaginary quadratic field is called its *class number*.

Ideal classes of $\mathbb{Z}[\sqrt{-5}]$. *The class number of* $\mathbb{Z}[\sqrt{-5}]$ *is 2.*

Proof. Let I be a nonprincipal ideal of $\mathbb{Z}[\sqrt{-5}]$ and let α, β be an integral basis of I constructed as in the proof of the lattice property of ideals in Section 11.6. That is, α is a nonzero element of minimal distance from 0, and β is of minimal distance from 0 among the elements of I not on the line through 0 and α.

Since I is an ideal, it includes the multiples $\alpha\sqrt{-5}$ and $\alpha(1+\sqrt{-5})$ of α, and 0. These four points together form the rectangle shown in Figure 12.2, a typical one in the principal ideal generated by α. (For the sake of simplicity, the line through 0 and α is shown horizontal in the diagram, but it is not necessarily the real axis.)

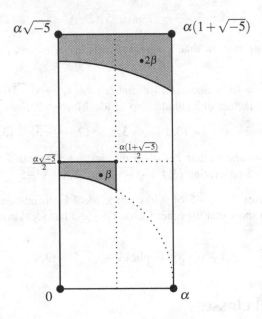

Figure 12.2: A rectangle of multiples of α

Since I is not the principal ideal (α), there are other members of I inside the rectangle, and hence in at least one of the quarters shown. Without loss of generality, we can assume that β lies in the bottom left quarter (if necessary, taking the difference between an element of I elsewhere in the rectangle and the nearest corner, or replacing α by $-\alpha$). Finally, since β is no nearer to 0 than α, by construction of the integral basis in Section 11.6, it lies in the shaded region of the bottom left quarter.

But then 2β lies in the shaded region in the top half of the rectangle, each point of which is clearly at distance $< |\alpha|$ from either $\alpha(1+\sqrt{-5})$ or $\alpha\sqrt{-5}$. This implies that the element $\alpha(1+\sqrt{-5}) - 2\beta$ or $\alpha\sqrt{-5} - 2\beta$ of I has absolute value $< |\alpha|$, contrary to the choice of α, unless

$$\beta = \alpha\frac{1+\sqrt{-5}}{2} \quad \text{or} \quad \beta = \alpha\frac{\sqrt{-5}}{2}.$$

In the first case, I has the same shape as the ideal $(2, 1 + \sqrt{-5})$, and hence belongs to the same class. The second case is impossible, because

$$\alpha \frac{\sqrt{-5}}{2} \in I \Rightarrow -\frac{5\alpha}{2} \in I, \quad \text{multiplying by } \sqrt{-5},$$

$$\Rightarrow \frac{\alpha}{2} \in I, \quad \text{adding } 3\alpha,$$

contrary to the choice of α as an element of minimal absolute value.   □

Exercises

The argument used above can also be used for $\mathbb{Z}[\sqrt{-6}]$, but it takes an interesting turn. The first possibility for β does not produce a nonprincipal ideal, but the second possibility does, so $\mathbb{Z}[\sqrt{-6}]$ has class number 2.

12.7.1 Show that $I = (2, 1 + \sqrt{-6}) = \{2m + (1 + \sqrt{-6})n : m, n \in \mathbb{Z}\}$ includes 1 and hence is $\mathbb{Z}[\sqrt{-6}]$, but that $J = (2, \sqrt{-6}) = \{2m + \sqrt{-6}n : m, n \in \mathbb{Z}\}$ is a nonprincipal ideal.

12.7.2 Rerun the argument above, explaining why only the second possibility for β now produces a nonprincipal ideal, so that the class number of $\mathbb{Z}[\sqrt{-6}]$ is 2 and that all nonprincipal ideals of $\mathbb{Z}[\sqrt{-6}]$ are of the form $(\alpha, \alpha\sqrt{-6}/2)$.

Returning to $\mathbb{Z}[\sqrt{-5}]$, we know that the nonprincipal ideal $(3, 1 + \sqrt{-5})$ is in the same class as $(2, 1 + \sqrt{-5})$, because we checked that it was the same shape in the exercises to Section 11.7. However, the integral basis $3, 1 + \sqrt{-5}$ of this ideal is not of the form $\alpha, \alpha(1 + \sqrt{-5})/2$.

12.7.3 Which integral basis of $(3, 1 + \sqrt{-5})$ is obtained by the process in the proof above?

12.8   Primes of the form $x^2 + 5y^2$

Now at last we know enough about $\mathbb{Z}[\sqrt{-5}]$ to be able to deal with the quadratic form $x^2 + 5y^2$ and the primes it represents. First observe what we can do with classical tools—congruences and quadratic reciprocity.

- Experience with the forms $x^2 + ny^2$ for $n = 1, 2, 3$ (Section 9.1) leads us to consider the values of $x^2 + 5y^2$ mod 20. The possible values of x^2 mod 20 are 1, 4, 9, 16, 5, and 0, hence the possible values of $5y^2$ mod 20 are 5 and 0. Prime values of $x^2 + 5y^2$ are odd and not divisible by 5, hence the possible prime values of $x^2 + 5y^2$ mod 20 are 1 and 9. That is, *primes of the form $x^2 + 5y^2$ are of the form $20n + 1$ or $20n + 9$.*

- To deal with $x^2 + 5y^2$ we likewise expect to need the quadratic character of -5. When $p = 20n + 1$ or $20n + 9$,

$$\left(\frac{-5}{p}\right) = \left(\frac{-1}{p}\right)\left(\frac{5}{p}\right) = 1 \times \left(\frac{p}{5}\right) \quad \text{by quadratic reciprocity}$$

$$= 1 \times 1$$

since $20n+1 \equiv 1 \equiv 1^2 \pmod 5$ and $20n+9 \equiv 4 \equiv 2^2 \pmod 5$. Hence -5 *is a square mod p when* $p = 20n+1$ *or* $p = 20n+9$.

We need prime ideal factorization to prove the converse of the first observation, namely, that every prime of the form $20n + 1$ or $20n + 9$ is of the form $x^2 + 5y^2$. Apart from the appearance of nonprincipal ideals, the proof is similar to the one for $x^2 + y^2$ in Chapter 6.

Primes of the form $x^2 + 5y^2$. *The primes of the form $x^2 + 5y^2$ are precisely those of the form $20n + 1$ or $20n + 9$.*

Proof. It remains to show that primes of the form $20n + 1$ or $20n + 9$ are of the form $x^2 + 5y^2$.

The second observation above shows that -5 is a square, mod p, for the primes $p = 20n + 1$ and $20n + 9$. In other words, for each such p there is an $m \in \mathbb{Z}$ such that

$$p \text{ divides } m^2 + 5 = (m + \sqrt{-5})(m - \sqrt{-5}).$$

However, p does *not* divide either factor $m + \sqrt{-5}$ or $m - \sqrt{-5}$ in $\mathbb{Z}[\sqrt{-5}]$, since $\frac{m}{p} \pm \frac{\sqrt{-5}}{p} \notin \mathbb{Z}[\sqrt{-5}]$. By unique prime ideal factorization in $\mathbb{Z}[\sqrt{-5}]$, it follows that (p) *is not a prime ideal of* $\mathbb{Z}[\sqrt{-5}]$, and hence it has a nontrivial prime ideal factorization.

The easy case is where one factor is a principal ideal, say $(a + b\sqrt{-5})$. Here we can argue as we did in $\mathbb{Z}[i]$ back in Section 6.3:

$$(p) = (a + b\sqrt{-5})C \quad \text{for some nontrivial ideal } C,$$

hence, taking conjugates,

$$(p) = (a - b\sqrt{-5})\overline{C} \quad \text{since } \overline{p} = p.$$

Multiplying the last two equations gives

$$(p^2) = (a^2 + 5b^2)C\overline{C} = (a^2 + 5b^2)(k),$$

for some $k \in \mathbb{Z}$, evaluating the product $C\overline{C}$ of the conjugate ideals as in Section 12.4. But then

$$p^2 = (a^2 + 5b^2)k \quad \text{is a nontrivial factorization in } \mathbb{Z},$$

and therefore $p = a^2 + 5b^2$.

The harder case is where all prime factors of (p) are nonprincipal ideals, and hence of the form $(\alpha, \alpha(1 + \sqrt{-5})/2)$ for some $\alpha \in \mathbb{Z}[\sqrt{-5}]$, by the previous section. Suppose

$$(p) = (\alpha, \alpha(1 + \sqrt{-5})/2)C.$$

Taking conjugates again, we obtain

$$(p) = (\overline{\alpha}, \overline{\alpha}(1 - \sqrt{-5})/2)\overline{C},$$

and multiplying the last two equations:

$$(p^2) = (\alpha, \alpha(1 + \sqrt{-5})/2)(\overline{\alpha}, \overline{\alpha}(1 - \sqrt{-5})/2)C\overline{C}.$$

Now $(\alpha, \alpha(1 + \sqrt{-5})/2)(\overline{\alpha}, \overline{\alpha}(1 - \sqrt{-5})/2) = (\alpha\overline{\alpha}/2)$, since its generators are $\alpha\overline{\alpha}$ and $3\alpha\overline{\alpha}/2$. And $C\overline{C} = (k)$ as before, so we have

$$2p^2 = \alpha\overline{\alpha} \cdot k = (a^2 + 5b^2)k \quad \text{for some } a, b, k \in \mathbb{Z}.$$

It follows that $p = a^2 + 5b^2$ or else $2p = a^2 + 5b^2$, for some $a, b \in \mathbb{Z}$.

The latter possibility is ruled out by congruences mod 20. From the values of x^2 and $5y^2$ mod 20 found above, we see that the possible even values of $a^2 + 5b^2$ mod 20 are 4, 6, 10, 14, 16. None of these match the values of $2p$, namely $40n + 2$ or $40n + 18$. Hence in all cases the prime $p = 20n + 1$ or $20n + 9$ is of the form $x^2 + 5y^2$.   □

Exercises

We have now classified the primes of the forms $x^2 + y^2$, $x^2 + 2y^2$, $x^2 + 3y^2$, and $x^2 + 5y^2$. Why did we skip the form $x^2 + 4y^2$?

12.8.1 Show that the primes of the form $x^2 + 4y^2$ are the same as those of the form $x^2 + y^2$, with one exception.

Primes of the form $x^2 + 6y^2$ can be found in much the same way as those of the form $x^2 + 5y^2$. For the hard step (where the prime ideal factors of (p) are all nonprincipal) we use the determination of nonprincipal ideals of $\mathbb{Z}[\sqrt{-6}]$ from the previous exercise set.

12.8.2 Use congruences mod 24 to show that any prime of the form $x^2 + 6y^2$ is of the form $24n + 1$ or $24n + 7$.

12.8.3 Use quadratic reciprocity to show that, for any prime p of the form $24n + 1$ or $24n + 7$, -6 is a square mod p.

12.8.4 Deduce from Exercise 12.8.3 that (p) is not a prime ideal of $\mathbb{Z}[\sqrt{-6}]$ when p is a prime of the form $24n + 1$ or $24n + 7$.

12.8.5 When (p) has a prime ideal factor of the form $(a + b\sqrt{-6})$, show that $p = a^2 + 6b^2$.

12.8.6 When all prime ideal factors of (p) are nonprincipal, hence of the shape $(\alpha, \alpha\sqrt{-6}/2)$ by Exercise 12.7.2, show that p has the form $x^2 + 6y^2$ in this case also.

12.9   Discussion

The treatment of rings, ideals, and quotient rings at the beginning of this chapter is probably the minimum required for significant applications to number theory. Along with some group theory and field theory, as found in *Elements of Algebra*, it should give enough algebraic background to read classical treatments of algebraic number theory, such as Hecke (1981). There one will find the theorem on unique prime factorization for the ideals of an arbitrary number field $\mathbb{Q}(\theta)$, where θ is an algebraic number, and the finiteness of the class number. These two theorems were first proved by Dedekind (1871), and his exposition of them in Dedekind (1877) is still worth reading, though not quite as streamlined as modern accounts. The finiteness of the class number, in particular, is usually proved today with the help of Minkowski's geometry of numbers.

The class number has a long history, beginning with the Lagrange (1773) idea of reducing binary quadratic forms. In the case of positive determinant, Lagrange gave an algorithm that finds the "simplest" equivalent of a given form, and at the same time finds a complete list of inequivalent forms. Thus the algorithm determines the *number* of inequivalent forms with given determinant D, in other words, the class number for D.

Gauss (1801) extended Lagrange's algorithm to binary quadratic forms with negative determinant, exploiting the inherent periodicity of such forms that we observed in Chapter 5. Finding a *formula* for the class number was much more difficult, and it was a turning point in the history of number theory when Dirichlet (1839) succeeded. His method uses L-series, a sophisticated generalization of the ζ-function discovered by Euler (and mentioned in Section 2.9). No substantially simpler approach has ever been found, and indeed L-series seem to be the right tool for the job. Dirichlet also used them to prove his theorem on primes in arithmetic progressions (see Section 9.9 for the background to this theorem). Both the class number formula and the theorem on primes may be found in Dirichlet's lectures on number theory, Dirichlet (1863), available in English translation.

Other approaches to the class number involve equally sophisticated mathematics, such as *modular functions*. The classical modular function $j(z)$ is a function defined on the upper half plane of $\mathbb{C}$, with the periodicity indicated by Figure 12.3, the *modular tessellation*.

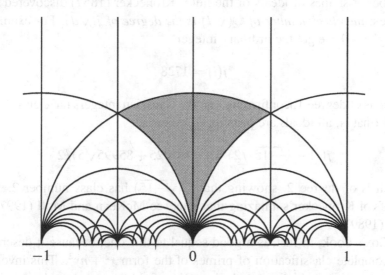

Figure 12.3: The modular tessellation

The function j maps the shaded region one-to-one onto $\mathbb{C}$ (one half onto the upper half plane, and the other half onto the lower), and repeats its values in every other region of the tessellation. The precise way to say this is that if $a, b, c, d \in \mathbb{Z}$ and $ad - bc = \pm 1$, then

$$j\left(\frac{az+b}{cz+d}\right) = j(z).$$

This periodicity property allows j to be viewed as a *function of lattice shapes*, as we now explain.

A lattice generated by $\omega_1, \omega_2 \in \mathbb{C}$ has "shape" given by the complex number ω_1/ω_2, because $|\omega_1/\omega_2| = |\omega_1|/|\omega_2|$ is the ratio of side lengths of a generating parallelogram, while $\arg \omega_1/\omega_2$ is the angle between the sides. However, the same lattice is generated by $a\omega_1 + b\omega_2$, $c\omega_1 + d\omega_2$ for any $a, b, c, d \in \mathbb{Z}$ with $ad - bc = \pm 1$, and hence its "shape" is represented as well by the number $(a\omega_1 + b\omega_2)/(c\omega_1 + d\omega_2) = (a\frac{\omega_1}{\omega_2} + b)/(c\frac{\omega_1}{\omega_2} + d)$. The shaded region of the modular tessellation contains exactly one representative of each lattice shape, so j gives each shape a different value, but (by periodicity) j takes the same value at each representative of a lattice shape.

Because of this, j may have something to say about the class number of imaginary quadratic fields, which (as we saw in Section 12.7) is the number of shapes of ideals of the field. Kronecker (1857) discovered that it does: *the class number of $\mathbb{Q}(\sqrt{d})$ is the degree of $j(\sqrt{d})$*. For example, with $d = -1$ we get the ordinary integer

$$j(i) = 1728$$

which is of degree 1, confirming that the Gaussian integers have class number 1 (that is, all ideals are principal). Likewise

$$j((1 + \sqrt{-15})/2) = (-191025 + 85995\sqrt{5})/2$$

which is of degree 2, showing that $\mathbb{Q}(\sqrt{-15})$ has class number 2. For proofs of Kronecker's amazing theorem, see McKean and Moll (1997) or Cox (1989).

Cox's book, in fact, is a good sequel to this one, because it describes the complete classification of primes of the form $x^2 + ny^2$. This involves not only more sophisticated algebra (class field theory), but also modular functions and related topics from analysis. Another book worth mentioning is Scharlau and Opolka (1985). As its title says, it covers the development of number theory "from Fermat to Minkowski", with particular emphasis on quadratic forms. It is in some ways complementary to this one, since it says little about ideal theory but is strong on analysis and the geometry of numbers. The latter topics are essential for anyone who wants to master number theory beyond the elements covered here.

Bibliography

Argand, J. R. (1806). *Essai sur un manière de représenter les quantités imaginaires dans les constructions géométriques*. Paris.

Artmann, B. (1999). *Euclid—the Creation of Mathematics*. Springer-Verlag, New York.

Baker, A. (1984). *A Concise Introduction to the Theory of Numbers*. Cambridge University Press, Cambridge.

Brouncker, W. (1657). Letter to Wallis, 3/13 October 1657. Translation in Fermat *Ouevres* 3: 419–420.

Cayley, A. (1858). A memoir on the theory of matrices. *Phil. Trans. Roy. Soc. London*, **148**, 17–37. In his *Collected Mathematical Papers* 2: 475–496.

Conway, J. H. (1997). *The Sensual (Quadratic) Form*. Mathematical Association of America, Washington, DC. With the assistance of Francis Y. C. Fung.

Cox, D. A. (1989). *Primes of the Form $x^2 + ny^2$*. John Wiley & Sons Inc., New York.

Coxeter, H. S. M. (1948). *Regular Polytopes*. Methuen & Co. Ltd., London.

Davenport, H. (1960). *The Higher Arithmetic: An Introduction to the Theory of Numbers*. Harper & Brothers, New York.

Dedekind, R. (1857). Abriss einer Theorie der höheren Kongruenzen in bezug auf einen reellen Primzahl-Modulus. *J. reine angew. Math.*, **54**, 1–26. In his *Werke* I: 40–67.

Dedekind, R. (1871). Supplement X. In Dirichlet's *Vorlesungen über Zahlentheorie*, 2nd ed., Vieweg 1871.

Dedekind, R. (1877). *Theory of Algebraic Integers*. Cambridge University Press, Cambridge. Translated from the 1877 French original and with an introduction by John Stillwell, 1996.

Dedekind, R. (1888). *Was sind und was sollen die Zahlen?* Braunschweig. English translation in *Essays on the Theory of Numbers*, Open Court, Chicago, 1901.

Dedekind, R. (1894). Supplement XI. In 4th edition of Dirichlet's *Vorlesungen über Zahlentheorie*.

Diffie, W. and Hellman, M. E. (1976). New directions in cryptography. *IEEE Trans. Information Theory*, **IT-22**(6), 644–654.

Dirichlet, P. G. L. (1837). Beweis des Satzes, dass jede unbegrentze arithmetische Progression, deren erstes Glied und Differenz ganze Zahlen ohne gemeinschaftlichen Factor sind, unendliche viele Primzahlen enthält. *Abh. Akad. Wiss. Berlin*, pages 45–81. In his *Werke* 1: 315–342.

Dirichlet, P. G. L. (1839). Recherches sur diverses applications de l'analyse infinitésimal à la théorie des nombres. *J. reine angew. Math.*, **19**, 324–369. In his *Werke* 1: 411–496.

Dirichlet, P. G. L. (1863). *Vorlesungen über Zahlentheorie*. F. Vieweg und Sohn, Braunschweig. English translation *Lectures on Number Theory*, with Supplements by R. Dedekind, translated from the German and with an introduction by John Stillwell, American Mathematical Society, Providence, RI, 1999.

Ebbinghaus, H.-D., Hermes, H., Hirzebruch, F., Koecher, M., Mainzer, K., Neukirch, J., Prestel, A., and Remmert, R. (1991). *Numbers*. Springer-Verlag, New York. With an introduction by K. Lamotke, Translated from the second 1988 German edition by H. L. S. Orde, Translation edited and with a preface by J. H. Ewing, Readings in Mathematics.

Eisenstein, F. G. (1844). Beweis des Reciprocitätssatzees für die cubischen Reste in der Theorie der aus dritten Wurzel der Einheit zusammengesetzten complexen Zahlen. *J. reine angew. Math.*, **27**, 289–310. Also in his *Mathematische Werke* 1: 59–80.

Euler, L. (1744). Theoremata circa divisores numerorum in hac forma $paa \pm qbb$ contentorum. *Comm. acad. sci. Petrop.*, **14**, 151–181. Also in his *Opera Omnia* ser. 1, vol.2, 194–222.

Euler, L. (1747). Letter to Goldbach, 6 May, 1747. In Fuss (1968), I, 413–420.

Euler, L. (1748a). *Introductio in analysin infinitorum, I*. Volume 8 of his *Opera Omnia*, series 1. English translation, *Introduction to the Analysis of the Infinite. Book I*, Springer-Verlag, 1988.

Euler, L. (1748b). Letter to Goldbach, 4 May 1748. In Fuss (1968), I, 450–455.

Euler, L. (1755). Demonstratio theorematis Fermatiani omnem numerum primum formae $4n + 1$ esse summam duorum quadratorum. *Novi comm. acad. sci. Petrop.*, **5**, 13–58. Also in his *Opera Omnia* ser. 1, vol. 2, 338–372.

Euler, L. (1756). Solutio generalis quorundam problematum diophanteorum quae vulgo nonnisi solutiones speciales admittere videntur. *Novi. comm. acad. sci. Petrop.*, **6**, 155–184. Also in his *Opera Omnia* ser. 1, vol. 2, 428–458.

Euler, L. (1770). *Elements of Algebra*. Translated from the German by John Hewlett. Reprint of the 1840 edition, with an introduction by C. Truesdell, Springer-Verlag, New York, 1984.

Fibonacci (1202). *Fibonacci's Liber Abaci*. Springer-Verlag. English translation by Laurence Sigler, 2002.

Fraenkel, A. (1914). Über die Teiler der Null und die Zerlegung von Ringen. *J. reine angew. Math.*, **145**, 139–176.

Fuss, P.-H. (1968). *Correspondance mathématique et physique de quelques célèbres géomètres du XVIIIème siècle. Tomes I, II.* Johnson Reprint Corp., New York. Reprint of the Euler correspondence originally published by l'Académie Impériale des Sciences de Saint-Pétersbourg. The Sources of Science, No. 35.

Gauss, C. F. (1801). *Disquisitiones arithmeticae*. Translated and with a preface by Arthur A. Clarke. Revised by William C. Waterhouse, Cornelius Greither and A. W. Grootendorst and with a preface by Waterhouse, Springer-Verlag, New York, 1986.

Gauss, C. F. (1832). Theoria residuorum biquadraticorum. *Comm. Soc. Reg. Sci. Gött. Rec.*, **4**. Also in his *Werke* 2: 67–148.

Genocchi, A. (1876). Généralisation du théorème de Lamé sur l'impossibilité de l'équation $x^7 + y^7 + z^7 = 0$. *C. R. Acad. Sci. Paris*, **82**, 910–913.

Graham, R. L., Knuth, D. E., and Patashnik, O. (1994). *Concrete Mathematics*. Addison-Wesley Publishing Company, Reading, MA, second edition.

Grassmann, H. (1861). Stücke aus dem Lehrbuche der Arithmetik. *Hermann Grassmann's Mathematische und Physikalische Werke*, **II/1**, 295–349.

Grosswald, E. (1966). *Topics from the Theory of Numbers*. The Macmillan Co., New York.

Hardy, G. H. (1937). The Indian Mathematician Ramanujan. *Amer. Math. Monthly*, **44**, 137–155.

Hardy, G. H. and Wright, E. M. (1979). *An Introduction to the Theory of Numbers*. The Clarendon Press Oxford University Press, New York, fifth edition.

Hasse, H. (1923). Über die Darstellbarkeit von Zahlen durch quadratische Formen im Körper der rationalen Zahlen. *J. reine angew. Math.*, **152**, 129–148.

Hecke, E. (1981). *Lectures on the Theory of Algebraic Numbers*. Springer-Verlag, New York. Translated from the German by George U. Brauer, Jay R. Goldman and R. Kotzen.

Ireland, K. F. and Rosen, M. I. (1982). *A Classical Introduction to Modern Number Theory*. Springer-Verlag, New York.

Kronecker, L. (1857). Über die elliptischen Functionen für welche complexe Multiplication stattfindet. In his *Werke* 4: 179–183.

Kummer, E. E. (1844). De numeris complexis, qui radicibus unitatis et numeris realibus constant. *Gratulationschrift der Univ. Breslau zur Jubelfeier der Univ. Königsberg*. Also in his *Collected Papers* 1: 165–192.

Lagrange, J. L. (1768). Solution d'un problème d'arithmétique. *Miscellanea Taurinensia*, **4**, 19ff. In his *Oeuvres* 1: 671–731.

Lagrange, J. L. (1770). Demonstration d'un théorème d'arithmétique. *Nouv. Mém. Acad. Berlin*. In his *Oeuvres* 3: 189–201.

Lagrange, J. L. (1773). Recherches d'arithmétique. *Nouv. mém. de l'acad. sci. Berlin*, page 265ff. Also in his *Oeuvres* 3: 695–795.

Lamé, G. (1847). Démonstration général du théorème de Fermat. *C. R. Acad. Sci. Paris*, **24**, 310–315.

Lebesgue, V. A. (1840). Démonstration de l'impossibilité de résoudre l'équation $x^7 + y^7 + z^7 = 0$ en nombres entiers. *J. Math. Pures Appl.*, **5**, 276–279.

Legendre, A.-M. (1785). Recherches d'analyse indéterminée. *Hist. de l'Acad. Roy. des Sci.*, pages 465–559.

Lemmermeyer, F. (2000). *Reciprocity Laws. From Euler to Eisenstein*. Springer-Verlag, Berlin.

Lenstra, Jr., H. W. (2002). Solving the Pell equation. *Notices Amer. Math. Soc.*, **49**(2), 182–192.

McKean, H. and Moll, V. (1997). *Elliptic Curves*. Cambridge University Press, Cambridge.

Mordell, L. J. (1969). *Diophantine Equations*. Academic Press, London.

Nagell, T. (1951). *Introduction to Number Theory*. John Wiley & Sons Inc., New York.

Peirce, B. (1881). Linear associative algebra. *Amer. J. Math.*, **4**, 97–229.

Pieper, H. (1978). *Variationen über ein zahlentheoretisches Thema von Carl Friedrich Gauß*. Birkhäuser Verlag, Basel. With a foreword by Hans Reichardt.

Rademacher, H. (1983). *Higher Mathematics from an Elementary Point of View*. Birkhäuser Boston, Mass. Edited by D. Goldfeld, With notes by G. Crane.

Redmond, D. (1996). *Number Theory*. Marcel Dekker Inc., New York.

Ribenboim, P. (1999). *Fermat's Last Theorem for Amateurs*. Springer-Verlag, New York.

Rivest, R. L., Shamir, A., and Adleman, L. (1978). A method for obtaining digital signatures and public-key cryptosystems. *Comm. ACM*, **21**(2), 120–126.

Rousseau, G. (1991). On the quadratic reciprocity law. *J. Austral. Math. Soc. Ser. A*, **51**(3), 423–425.

Samuel, P. (1970). *Algebraic Theory of Numbers*. Houghton Mifflin Co., Boston, MA.

Scharlau, W. and Opolka, H. (1985). *From Fermat to Minkowski*. Springer-Verlag, New York. Translated from the German by Walter Kauffmann-Bühler and Gary Cornell.

Shor, P. W. (1994). Algorithms for quantum computation: discrete logarithms and factoring. In *35th Annual Symposium on Foundations of Computer Science (Santa Fe, NM, 1994)*, pages 124–134. IEEE Comput. Soc. Press, Los Alamitos, CA.

Silverman, J. H. and Tate, J. (1992). *Rational Points on Elliptic Curves*. Springer-Verlag, New York.

Smith, D. E. (1959). *A Source Book in Mathematics*. Dover Publications Inc., New York. 2 vols.

Stillwell, J. (1994). *Elements of Algebra*. Springer-Verlag, New York.

Stillwell, J. (1998). *Numbers and Geometry*. Springer-Verlag, New York.

Viète, F. (1593). Variorum de rebus mathematicis responsorum libri octo. In his *Opera*, 347–435.

Weil, A. (1984). *Number Theory. An Approach through History, from Hammurapi to Legendre*. Birkhäuser Boston Inc., Boston, MA.

Wessel, C. (1797). Om Directionens analytiske Betegning, et Forsøg anvendt fornemmelig til plane og sphæriske Polygoners Opløsning. *Danske Selsk. Skr. N. Samml.*, **5**. English translation in Smith (1959), vol. 1, 55–66.

Yan, S. Y. (2000). *Number Theory for Computing*. Springer-Verlag, Berlin.

Index

additive inverse
  property, 182
Adleman, 68
al-Haytham, 53
al-Khazin, 18
algorithm
  continued fraction, 26
  Euclidean, 22
    and RSA, 66
  for binary numeral, 8
    and RSA, 66
  for class number, 236
  for Egyptian fractions, 4
  for quadratic characters, 170
Archimedes' "cattle problem", 99
Argand, 155
arithmetic progression
  primes in, 178, 237
  rule, 96
associate, 183
associative law, 182
  for +, 182
  for ×, 182

Babylonian numerals, 13
Baker, 193
Bhaskara II, 84
  and $x^2 - 61y^2 = 1$, 100
binary notation, 1
  algorithm, 8
binomial theorem, 64
biquadratic
  character, 179
  reciprocity, 179
Brahmagupta, 82

composition, 82
identity, 219
solved $x^2 - 92y^2 = 1$, 99
Brouncker, 100

$\mathbb{C}$, 18
  matrix representation, 139
Caesar cipher, 67
cancellation
  in integral domain, 225
  of ideals, 229
casting out nines, 43, 44, 47
Cayley, 142, 194
Chinese remainder theorem, 158, 171
  and inverse mod ab, 174
  and phi function, 175
  and quadratic reciprocity, 176
  classical, 172
  full, 173
class
  field theory, 238
  group, 219
  number, 214, 218
    algorithm, 236
    finiteness, 236
    formula, 236, 237
    of $\mathbb{Z}[\sqrt{-5}]$, 231
    of $\mathbb{Z}[\sqrt{-6}]$, 233
    via modular function, 238
  of ideals, 219, 221, 231
Cocks, 75
commutative law, 182
  for +, 182
  for ×, 182
composition